Progress in Molecular and Subcellular Biology

Volume 61

Editor-in-Chief

Werner E. G. Müller, ERC Advanced Investigator Group, Institute for Physiological Chemistry, University Medical Center of the Johannes Gutenberg University, Mainz, Germany

Series Editors

Heinz C. Schröder, ERC Advanced Investigator Group, University Medical Center of the Johannes Gutenberg University, Mainz, Germany

Đurđica Ugarković, Rudjer Boskovic Institute, Zagreb, Croatia

This series gives an insight into the most current, cutting edge topics in molecular biology, including applications in biotechnology and molecular medicine. In the recent years, the progress of research in the frontier area of molecular and cell biology has resulted in an overwhelming amount of data on the structural components and molecular machineries of the cell and its organelles and the complexity of intra- and intercellular communication. The molecular basis of hereditary and acquired diseases is beginning to be unravelled, and profound new insights into development and evolutionary biology, as well as the genetically driven formation of 3D biological architectures, have been gained from molecular approaches. Topical volumes, written and edited by acknowledged experts in the field, present the most recent findings and their implications for future research. This series is indexed in PubMed.

Werner E. G. Müller • Heinz C. Schröder •
Patrick Suess • Xiaohong Wang
Editors

Inorganic Polyphosphates

From Basic Research to Medical Application

Springer

Editors
Werner E. G. Müller
ERC Advanced Investigator Group
Institute for Physiological Chemistry
University Medical Center of the Johannes Gutenberg University
Mainz, Germany

Heinz C. Schröder
ERC Advanced Investigator Group, Institute for Physiological Chemistry
University Medical Center of the Johannes Gutenberg University
Mainz, Germany

Patrick Suess
Department of Biological Chemistry
University of Michigan Medical School
Ann Arbor, MI, USA

Xiaohong Wang
ERC Advanced Investigator Group, Institute for Physiological Chemistry
University Medical Center of the Johannes Gutenberg University
Mainz, Germany

ISSN 0079-6484 ISSN 2197-8484 (electronic)
Progress in Molecular and Subcellular Biology
ISBN 978-3-031-01239-6 ISBN 978-3-031-01237-2 (eBook)
https://doi.org/10.1007/978-3-031-01237-2

This Springer imprint is published by the registered company Springer Nature Switzerland AG
The registered company address is: Gewerbestrasse 11, 6330 Cham, Switzerland

Preface

Research on the occurrence, metabolism, and biological function of inorganic polyphosphates (polyP) has experienced rapid development in recent years. Previous research mainly concerned polyP, which is found in yeast and bacteria. Only later, initiated by the work of Arthur Kornberg and contributors to this book, was it found that these polymers are also present in some significant amounts in higher eukaryotic organisms, including humans. For this purpose, new analytical methods for detecting and determining the size of these polymers had to be developed. More than 20 years ago, together with contributions of the protagonists in this field, Arthur Kornberg and Igor S. Kulaev, we have summarized the state of knowledge at that time about polyP and the enzymatic basis of their biosynthesis and degradation in Vol. 23 of this series (Schröder HC, Müller WEG, eds, Inorganic Polyphosphates—Biochemistry, Biology, Biotechnology Prog Mol Subcell Biol 23:1–317; 1999). In the following decade, the biomedical importance of these polymers was recognized, mainly due to the discovery of their morphogenetic effects and their potential for application in regenerative medicine, comparable only to another group of inorganic biomaterials, the biogenic silica (biosilica). These results along with other inorganic polymers of medical interest have been compiled in Vol. 54 of this series (Müller WEG, Wang XH, Schröder HC, eds, Biomedical Inorganic Polymers: Bioactivity and Applications of Natural and Synthetic Polymeric Inorganic Molecules. Prog Mol Subcell Biol 54:1–303; 2013). Now, in the last few years, pioneered by the work of Andrey Y. Abramov and Evgeny V. Pavlov as well as the ERC Investigator team of Werner E.G. Müller, together with his colleagues Xiaohong Wang and Heinz C. Schröder, it became clear that, in addition to its morphogenetic activity, polyP is characterized by another property that is unique among inorganic biopolymers: the storage and release of metabolic energy. There is no other biopolymer that is able to store as much chemical energy in the form of high-energy phosphoanhydride bonds as the longer-chain polyP polymers. This groundbreaking discovery gave an impetus for the development of therapeutic strategies to treat pathological conditions characterized by a lack of energy.

The main factories of energy in the cell, in the form of the universal energy carrier molecule ATP, are the mitochondria. These cell organelles have also been recognized as the main producers of energy-rich polyP, alongside the acidocalcisomes as the main storage organelles of the polymer. In Chap. 1 of his volume, Artyom Y. Baev and Andrey Y. Abramov review the role of mammalian mitochondrial F_0F_1-ATP synthase in the production of this polymer. The relationship between mitochondrial energy metabolism and stress-related mitochondrial pathologies is discussed in Chap. 2 by Maria A. Neginskaya and Evgeny V. Pavlov. In particular, the possible function of polyP in the transport of ions of mitochondria, including the regulation of calcium hemostasis and the mitochondrial permeability transition pore, is discussed. Chapter 3, written by Pedro Urquiza and Maria E. Solesio, focuses on the functional significance and disturbances of mitochondrial polyP metabolism in the development of neurodegenerative diseases and other age-dependent disorders. Another strongly energy-dependent mechanism is wound healing, and energy deficiency in various pathological conditions such as diabetes, cardiovascular diseases, or infections is a major factor impairing this process, as is the case in chronic wound healing. This lack of metabolic energy particularly affects the extracellular space, and administration of exogenous polyP as a source for extracellular ATP generation has shown promise as a suitable strategy in the therapy of non-healing or difficult-to-treat wounds, as discussed in Chap. 4 by Xiaohong Wang and coauthors. Metabolic energy is also required in the healing of bone and cartilage defects. The development and use of novel polyP-based materials that exploit their energy-delivering function in addition to their morphogenetic activity in regenerative bone and cartilage repair is discussed in Chap. 5 by Heinz C. Schröder and coauthors. PolyP has also been shown to have a regulatory function in the human immune system, as well as exhibiting a significant antiviral activity. In Chap. 6, Patrick M. Suess summarizes the effects of polyP on leukocytes and their role in inflammatory and immune responses. Recent results showed that polyP acts as an efficient inhibitor of binding of the SARS-CoV-2 spike protein to its cellular receptor protein. In Chap. 7, Werner E. G. Müller and coauthors discuss the underlying mechanism and other antiviral target sites of this unique and versatile bioinorganic polymer.

This book summarizes the latest advances in research on polyP in mammalian organisms and the exciting possibilities that this energy-delivering and regeneratively active polymer opens for therapy of human disorders.

Mainz, Germany — Werner E. G. Müller
Mainz, Germany — Heinz C. Schröder
Ann Arbor, MI — Patrick Suess
Mainz, Germany — Xiaohong Wang

Contents

1 **Inorganic Polyphosphate and F_0F_1-ATP Synthase of Mammalian Mitochondria** 1
Artyom Y. Baev and Andrey Y. Abramov

2 **Inorganic Polyphosphate in Mitochondrial Energy Metabolism and Pathology** 15
Maria A. Neginskaya and Evgeny V. Pavlov

3 **Inorganic Polyphosphate, Mitochondria, and Neurodegeneration** . . . 27
Pedro Urquiza and Maria E. Solesio

4 **Polyphosphate in Chronic Wound Healing: Restoration of Impaired Metabolic Energy State** 51
Xiaohong Wang, Hadrian Schepler, Meik Neufurth, Shunfeng Wang, Heinz C. Schröder, and Werner E. G. Müller

5 **Biomimetic Polyphosphate Materials: Toward Application in Regenerative Medicine** 83
Heinz C. Schröder, Xiaohong Wang, Meik Neufurth, Shunfeng Wang, and Werner E. G. Müller

6 **Effects of Polyphosphate on Leukocyte Function** 131
Patrick M. Suess

7 **Polyphosphate in Antiviral Protection: A Polyanionic Inorganic Polymer in the Fight Against Coronavirus SARS-CoV-2 Infection** 145
Werner E. G. Müller, Xiaohong Wang, Meik Neufurth, and Heinz C. Schröder

Contributors

Andrey Y. Abramov Department of Clinical and Movement Neurosciences, UCL Queen Square Institute of Neurology, London, UK

Artyom Y. Baev Laboratory of Experimental Biophysics, Centre for Advanced Technologies, Tashkent, Uzbekistan

Werner E. G. Müller ERC Advanced Investigator Group, Institute for Physiological Chemistry, University Medical Center of the Johannes Gutenberg University, Mainz, Germany

Maria A. Neginskaya Department of Molecular Pathobiology, New York University, New York, NY, USA

Meik Neufurth ERC Advanced Investigator Group, Institute for Physiological Chemistry, University Medical Center of the Johannes Gutenberg University, Mainz, Germany

Evgeny V. Pavlov Department of Molecular Pathobiology, New York University, College of Dentistry, New York, NY, USA

Hadrian Schepler Department of Dermatology, University Clinic Mainz, Mainz, Germany

Heinz C. Schröder ERC Advanced Investigator Group, Institute for Physiological Chemistry, University Medical Center of the Johannes Gutenberg University, Mainz, Germany

Maria E. Solesio Department of Biology, Rutgers University, Camden, NJ, USA

Patrick M. Suess Department of Biological Chemistry, University of Michigan Medical School, Ann Arbor, MI, USA

Pedro Urquiza Department of Biology, Rutgers University, Camden, NJ, USA

Shunfeng Wang ERC Advanced Investigator Group, Institute for Physiological Chemistry, University Medical Center of the Johannes Gutenberg University, Mainz, Germany

Xiaohong Wang ERC Advanced Investigator Group, Institute for Physiological Chemistry, University Medical Center of the Johannes Gutenberg University, Mainz, Germany

Chapter 1
Inorganic Polyphosphate and F_0F_1-ATP Synthase of Mammalian Mitochondria

Artyom Y. Baev and Andrey Y. Abramov

Abstract Inorganic polyphosphate is a polymer which plays multiple important roles in yeast and bacteria. In higher organisms the role of polyP has been intensively studied in last decades and involvements of this polymer in signal transduction, cell death mechanisms, energy production, and many other processes were demonstrated. In contrast to yeast and bacteria, where enzymes responsible for synthesis and hydrolysis of polyP were identified, in mammalian cells polyP clearly plays important role in physiology and pathology but enzymes responsible for synthesis of polyP or consumption of this polymer are still not identified. Here, we discuss the role of mitochondrial F_0F_1-ATP synthase in polyP synthesis with results, which confirm this proposal. We also discuss the role of other enzymes which may play important roles in polyP metabolism.

Keywords Mitochondria · Membrane potential · F_0F_1-ATP synthase · ADP/O ratio · Endopolyphosphatase · Alkaline phosphatase · Plasma membrane calcium pump · H-prune · Tartrate-resistant acid phosphatase · NAD kinase

Abbreviations

$\Delta\Psi_m$	Mitochondrial membrane potential
ADP	Adenosine-5′-diphosphate
AK	Adenylate kinase
ALP	Alkaline phosphatase
AMP	Adenosine-5′-monophosphate

A. Y. Baev
Laboratory of Experimental Biophysics, Centre for Advanced Technologies, Tashkent, Uzbekistan

A. Y. Abramov (✉)
Department of Clinical and Movement Neurosciences, UCL Queen Square Institute of Neurology, London, UK
e-mail: a.abramov@ucl.ac.uk

W. E. G. Müller et al. (eds.), *Inorganic Polyphosphates*, Progress in Molecular and Subcellular Biology 61, https://doi.org/10.1007/978-3-031-01237-2_1

ATP	Adenosine-5′-triphosphate
H-prune	Human metastasis regulator protein
P_i	Inorganic phosphate
PolyP	Inorganic polyphosphate
PPK	Polyphosphate kinase
PPX	Exopolyphosphatase
PTP	Permeability Transition Pore
TRAP	Tartrate-resistant acid phosphatase

1.1 Introduction

Inorganic polyphosphates (polyPs) are the ancient homopolymers, which consist of several to thousands of orthophosphates residues (Fig. 1.1). PolyP is found in all living creatures, however, the role of this polymer in the physiology and pathophysiology of some organisms including mammalians is still disputable. Thus, while its role in microorganisms is well studied and the enzymes responsible for the polyP metabolism discovered decades ago, investigation of polyP metabolism and functions in mammalian cells is still a cutting edge topic. Despite the fact that polyPs were firstly isolated from yeast at the end on nineteenth century, the main information regarding its functions in microorganisms and unicellular eukaryotes started to appear in the second part of twentieth century. The group of A. Kornberg made a big breakthrough in the polyP studies after isolation, purification, and cloning of the polyP kinase (PPK) from *Escherichia coli* and exopolyphosphatases (PPX) from *E. coli* (Ahn and Kornberg 1990; Akiyama et al. 1993) and *Saccharomyces cerevisiae* (Wurst and Kornberg 1994). PolyP kinase is responsible for reversible conversion of the terminal (γ) phosphate of ATP to polyP and vice versa. Exopolyphosphatases hydrolyze the terminal residues of polyP to P_i, however, while the *E. coli* PPX prefers long-chain polyPs, the exopolyphosphatase isolated from *S. cerevisiae* (scPPX1) can hydrolyze polyPs of different chain length. Moreover, scPPX1 is 40 times more effective in its activity compared to *E. coli* PPX (Kornberg 1999). This knowledge helped a lot in the creation of genetic tools, which

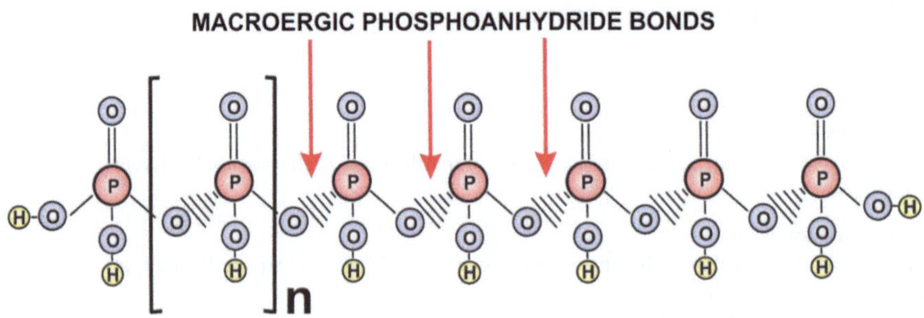

Fig. 1.1 Structure of inorganic polyphosphates

could regulate the level of polyPs in different cells. Many functions of polyP in mammalian cells were discovered by transfection of different mammalian cells with scPPX1 or PPK.

1.2 F_0F_1 ATP Synthase and PolyP

Recently we attempted to prove the hypothesis, which we proposed in 2010 (Pavlov et al. 2010), that mitochondrial F_0F_1 ATP synthase is one of the enzymes responsible for the synthesis of inorganic polyphosphates in mammalian cells (Baev et al. 2020).

Mitochondrial F_0F_1 ATP synthase is an enzyme, which uses proton gradient ($\Delta\Psi_m$), generated by the activity of the electron transport chain, for the synthesis of ATP. This unique enzyme can work in both directions—ATP synthesis, using the $\Delta\Psi_m$; and in reverse mode—using the energy of ATP hydrolysis for pumping protons from mitochondrial matrix to intermembrane space, when $\Delta\Psi_m$ is decreased (Fig. 1.2a, b) (Zorova et al. 2018). It is well known that mitochondria produce majority of cellular ATP, however, decrease of $\Delta\Psi_m$ can lead to the activation of cell death pathways, so during different pathologies, when the $\Delta\Psi_m$ decreases, mitochondria start to consume cellular ATP for the $\Delta\Psi_m$ maintenance (Chinopoulos and Adam-Vizi 2010). To prove that mitochondrial F_0F_1 ATP synthase is involved in the polyP metabolism we applied four different methods, two of which showed that this enzyme is able to synthesize polyP and two others that it can hydrolyze it.

One of the best methods to assure the quality of isolated mitochondria is to measure how they consume oxygen for oxidative phosphorylation purposes. We measured mitochondrial respiration with the help of Clark electrode. In our experiments, we evaluated different stages of mitochondrial respiration by Chance (Fig. 1.2c) (Chance and Williams 1955):

V_2—"passive" mitochondrial respiration, when there are substrates for electron transport chain (glutamate and malate) and PO_4^{3-} in the system but no ADP (substrate for oxidative phosphorylation); V_3—mitochondrial respiration in the presence of ADP, stage when oxidative phosphorylation take place; V_4—the stage when all added ADP converted into ATP and mitochondrial respiration rate reverting close to V_2 level. This parameter might be altered if the quality of mitochondrial isolation is not very good or during different pathological conditions; V_{CCCP}—the mitochondrial respiration rate in the presence of uncoupler CCCP. Usually, it is the maximal respiration rate as far as phosphorylation processes do not limit the work of electron transport chain. This parameter should be higher than V_3 or on the same level with it (Fig. 1.2). *RC*—respiratory coefficient—V_3/V_4 ratio. This parameter shows the coupling of oxidation and phosphorylation processes during mitochondrial respiration and varying from 3 to 15 in healthy mitochondria; *ADP/O*—the ratio of added ADP (converted by mitochondria to ATP) to the amount of oxygen used during the oxidative phosphorylation process (V_3). This parameter shows the efficiency of mitochondrial oxidative phosphorylation.

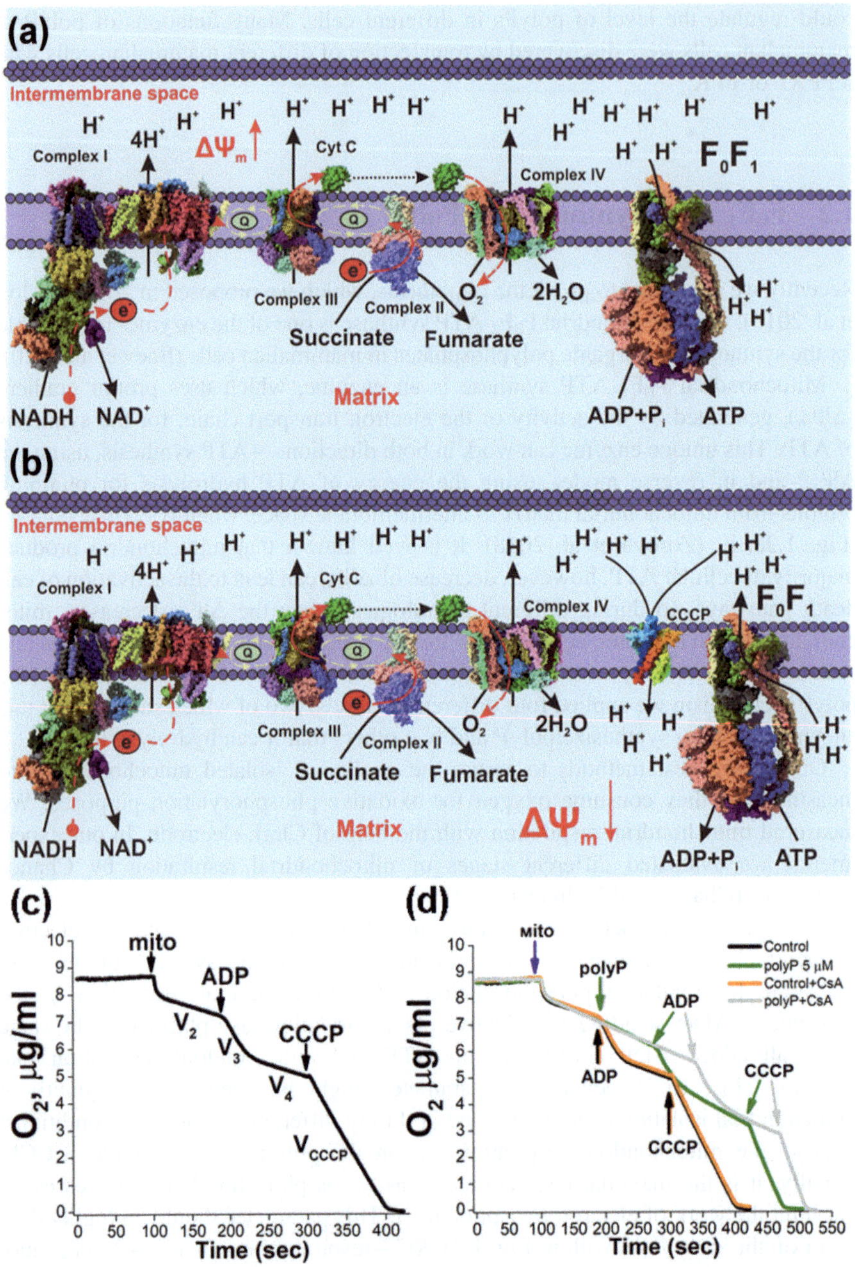

Fig. 1.2 Scheme of function of the respiratory chain. (**a**, **b**) Coupled and uncoupled states of oxidative phosphorylation; (**c**) States of mitochondrial respiration by Chance; (**d**) Classical inhibitor of PTP CsA does not prevent the effect of polyP on mitochondrial oxidative phosphorylation

We found that addition of polyP to the isolated mitochondria increased the mitochondrial ADP independent respiration (V_2). In contrast, presence of polyP suppressed ADP-dependent respiration and almost eliminated V_4 (the effect was concentration dependent). Usually, these results indicate mild uncoupling of mitochondria and for us, it was understandable as far as we previously showed that polyP in physiological concentration might modulate permeability of mitochondrial membranes through activation of mitochondrial permeability transition pore (PTP) (Baev et al. 2017). However, addition of CCCP increased the respiration rate in the presence of polyP the same as in control experiments, which indicated that in the presence of polyP mitochondria are in fully coupled state. To exclude the effect of polyP on mitochondrial PTP, we performed experiments with an inhibitor of PTP Cyclosporine A (CsA). The application of 1 μM of CsA did not change the respiration rate of control mitochondria and the effect of polyP on mitochondrial respiration (Fig. 1.2d). From the control experiments, it is perfectly seen that the addition of ADP—a substrate for oxidative phosphorylation and F_0F_1 ATP synthase increases respiration rate (V_3) (Fig. 1.2c). It is happening because F_0F_1 starts to synthesize ATP from ADP and P_i and, as we know, it requires the energy of $\Delta\Psi_m$ for this process. Protons from intermembrane space flow throw F_0 into mitochondrial matrix; its energy helps F_0F_1 to change the conformation and create more favorable environment for ATP synthesis (Fig. 1.2a). However, at the same time this event lowers $\Delta\Psi_{m,}$ which is the signal for electron transport chain (I–IV complexes) to work faster. For the experiments, we add a certain amount of ADP, so when F_0F_1 converts all added ADP into ATP the oxygen consumption decreases and we can see V_4 state. If the amount of added ADP is 5–10 times higher, the V_3 state would not reach V_4.

Addition of polyP increases mitochondrial respiration, the same as we can see after application of ADP, but the rate of respiration in the presence of ADP is much higher than in the presence of polyP. In classical scenario, F_0F_1 ATP synthase adds one P_i to the ADP and after that ATP leaves the active site of the enzyme. In case of polyP, we do not know how many inorganic phosphates the enzyme can add to the polymer, so the effect of polyP on oxygen consumption might be longer.

To verify that polyP might be produced by F_0F_1 ATP synthase we monitored the level of polyP in isolated mitochondria with the help of DAPI (Aschar-Sobbi et al. 2008). In 2010 Pavlov et al. showed that modulation of the activity of respiratory chain by different inhibitors and substrates can change the level of polyP in real time (Pavlov et al. 2010). In our experiments, we also showed that addition of glutamate and malate or succinate (substrates for I and II respiratory complexes, respectively) caused rise in the level of polyP, while inhibition of respiration by rotenone and oligomycin stopped polyP production. Measuring the level of ATP in the same conditions will show the same results—concentration rise in the presence of substrates for oxidative phosphorylation and downfall in the presence of inhibitors. Taking into account that in microorganisms ATP might be used for polyP production by polyphosphate kinase and vice versa, it was always a question if the rise of polyP level in mitochondria depends on previous production of ATP. Therefore, in our experiments, when we monitored the level of polyP in mitochondria, we performed two series of experiments: we added ATP at the beginning of the experiment to the system without substrates for oxidative phosphorylation; and at the end of the experiment, to the system with substrates for oxidative phosphorylation, but when the polyP production was stopped by oligomycin.

In both cases, the addition of ATP did not lead to the rise of polyP level, so we concluded that there is no intermediate enzyme in mitochondria, which uses ATP production during oxidative phosphorylation for polyP synthesis.

As we mentioned before when $\Delta\Psi_m$ is decreasing F_0F_1 ATP synthase starts to consume ATP. There are many ways how to measure this activity, the main point is to decrease or eliminate $\Delta\Psi_m$ to switch the activity of the enzyme into hydrolyzing mode. Most of the detection methods are based on the release of P_i from ATP during hydrolysis, but in our case because of the nature of polyP these methods could not be used as far as polyP signal would interfere with P_i signal. When P. Mitchell tried to prove his chemiosmotic theory, the main idea of which was that during electron transport mitochondria generate proton gradient in intermembrane space (Fig. 1.2a, b) and afterward uses it for phosphorylation purposes, he applied elegant method for detection of ATP hydrolysis by oligomycin sensitive mitochondrial ATP synthase (Mitchell and Moyle 1968). This method is based on the idea that ATP hydrolysis at pH values close to neutral leads to the appearance of protons and acidulation of the incubation media, which is detectable via pH electrode. For this series of experiments, we used submitochondrial particles. We showed that in the presence of polyP there was significant increase in acidulation rate, which was not dependent on the presence of ATP and was blocked by oligomycin. However, as far as we used submitochondrial particles with many other mitochondrial enzymes in the system present, in the next series of experiments we lysed mitochondria, and immunocaptured mitochondrial F_0F_1 ATPase in the microplate well via specific antibodies—so we created a system which contained just F_0F_1 ATPase, without other mitochondrial enzymes. Our experiments showed that F_0F_1 ATPase could hydrolyze polyP in oligomycin dependent manner.

1.2.1 Docking Experiments

To understand wether polyP can bind to the catalytic sites of ATP synthase we performed docking experiments. For this purpose, we used bovine F_0F_1 ATP synthase monomer in first rotational state (6ZPO) from protein data bank. The structure of this protein was determined by electron cryo-microscopy by John Walker's group and was published in 2020 (Spikes et al. 2020). For docking experiments we used polyP$_{14}$ (SpolyP), which has 14 residues of inorganic phosphate in the chain. The structure of SpolyP was created and minimized with the help of ChemOffice tools and converted into pdb format with the help of OpenBabel software (O'Boyle et al. 2011). The F_0F_1 ATP synthase (6ZPO) file was downloaded from protein data bank website in pdb format and was prepared for docking in UCSF Chimera (Pettersen et al. 2004). Docking experiments were performed with help of AutoDock Vina (Trott and Olson 2010). For the first round of docking, we decided

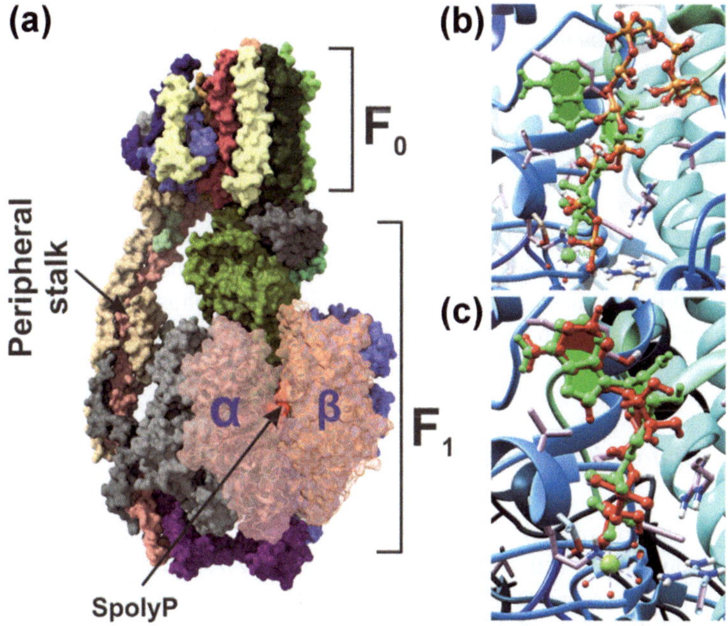

Fig. 1.3 Inorganic polyphosphate binds to the ADP binding site of F_1. (**a**) Position of the docked polyP on F_0F_1 ATP synthase; (**b**) PolyP (red-orange molecule) binds exactly to the ADP binding site of F1—comparison of docked polyP position with the position of ADP (green molecule)—ligand that was present in the original F_0F_1 ATP synthase molecule from protein data bank (6ZPO); (**c**) Position comparison of the docked ADP (red molecule) and ADP which was present in the original F_0F_1 ATP synthase molecule

Table 1.1 Binding energies of SpolyP and ADP

Ligand	$\Delta G_{binding}$ (kcal/mol)
SpolyP (F_1 region)	−9.3
SpolyP (ADP binding site)	−9.6
ADP (ADP binding site)	−9.2

to cover all F_1 region and to see all possible sites where polyP can bind to the enzyme. Our results showed that SpolyP selectively binds to the ADP binding site of F_1 region with Autodock Vina binding score −9.3 (Fig. 1.3a, Table 1.1). During the preparation of the enzyme for docking experiments, we deleted all ligands that were attached to the ATP synthase. After the docking with polyP, we merged our docking results with original structure of ATP synthase (6ZPO), which contained all ligands. It indicates that polyP binds exactly to ADP binding site (Fig. 1.3b). In the second round of docking, we narrowed the grid box and focused it to ADP binding site, which we had found in previous round. It slightly increased the binding energy values to −9.6 (Table 1.1). To understand that our calculations are adequate we performed docking experiments with ADP molecule. Once again, we merged our docking experiments with original structure of ATP synthase and saw that our docked molecule (Fig. 1.3c, red color) binds to the same site and at the same conformation as a ligand in the original ATP synthase (Fig. 1.3c, green color).

1.3 Mammalian Enzymes of PolyP Metabolism

Despite the fact that metabolism of polyP in mammalian cells is still an open question, there are many reports, which show that some of the mammalian enzymes can use polyP as a substrate for phosphorylation or can phosphorylate polyP itself (elongate it). We want to describe some of these enzymes (Fig. 1.4).

1.3.1 Endopolyphosphatase

In 1996 group of A. Kornberg showed that mammalian tissues have endopolyphosphatase activity (Kumble and Kornberg 1996). In this research, vacuolar endopolyphosphatase from *S. cerevisiae* was extracted, purified and its activity regarding polyP was tested. Moreover, in this research crude extracts from mammalian brain, heart, kidney, lungs, and liver was tested on endopolyphosphatase activity, the highest endopolyphosphatase activity was present in brain tissues (Kumble and Kornberg 1996). Endopolyphosphatases are the class of enzymes that are also called polyP depolymerases, they cleave long-chain polyphosphates into short and medium chains. These types of enzymes were found in many

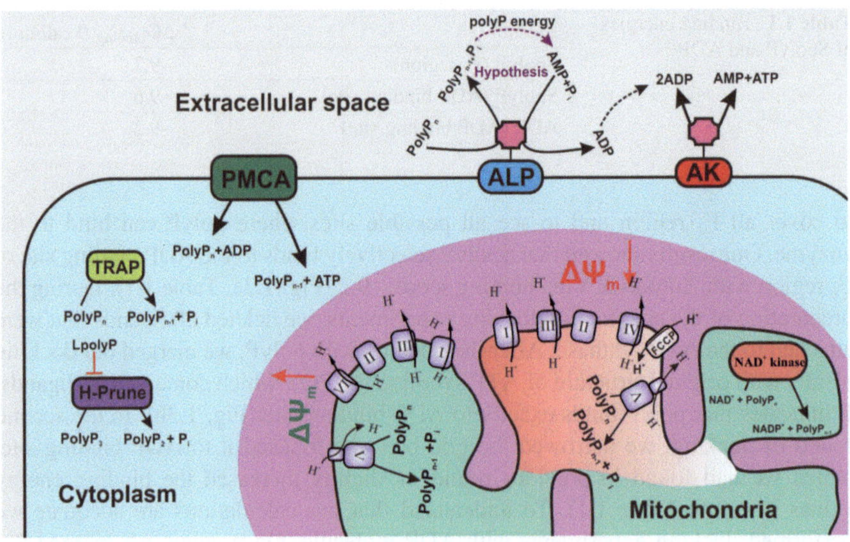

Fig. 1.4 Mammalian enzymes, which can produce inorganic polyphosphate or use it as a substrate

microorganisms and unicellular eukaryotes. Despite the fact that mammalian tissues showed endopolyphosphatase activity, alignment of PPN1 sequence of *S. cerevisiae* with mammalian genome has not showed any similarities. Considering the fact that in this particular research, the group worked with crude extracts, no other research groups succeeded to extract and purify the endopolyphosphatase from mammalian tissues and whole genome alignment has not showed any similarities between mammalian genes and PPN1 of *S. cerevisiae* we believe it is rather hasty to conclude that mammals have endopolyphosphatases. However, considering the recent report, that bacterial long chain polyphosphates can disturb multiple functions in mammalian cells (Roewe et al. 2020) we believe that mammalian cells should have an enzyme which could cleave long chain polyphosphates, but probably it is not an endopolyphosphatase but another enzyme with endopolyphosphatase activity.

1.3.2 PMCA of Erythrocytes

Reusch et al. in 1997 showed plasma membrane calcium pump (PMCA) isolated from human erythrocytes is associated with two homopolymers—inorganic polyphosphate and poly-3-hydroxybutyrate (PHB) (Reusch et al. 1997). This tandem of polymers was also found in other transporting systems. First of all these polymers can themselves form voltage-activated calcium channels in the plasma membrane of different bacteria (Reusch et al. 1995). Further studies had shown that this complex does not have any proteins, which is currently the only case of an ion channel of "non-protein" origin (Das et al. 1997). In 2005, a similar complex was extracted from mammalian mitochondria. Incorporation of this complex into planar lipid bilayer showed that it possess biophysical and electrophysiological properties of mitochondrial Permeability Transition Pore (PTP) (Pavlov et al. 2005). In the period from 2007 to 2009, several studies revealed that polyP and PHB play crucial role in functioning of KcsA (Negoda et al. 2007, 2009), TRPA1 (Kim and Cavanaugh 2007), and TRPM8 (Zakharian et al. 2009) channels, in KcsA and TRPM8, polyP, along with PHB, plays role as a structural component of the channels. In 2016 group of E. Pavlov showed that polyP and PHB have an association with subunit c of F_0F_1 ATP synthase, which is believed a main structural part of mitochondrial PTP (Elustondo et al. 2016; Solesio et al. 2016).

Furthermore, Reusch et al. showed in their research that PMCA of human erythrocytes can transfer orthophosphate residues from ATP to polyP and vice versa, hereby showing polyphosphate kinase activity (Reusch et al. 1997). PMCA belongs to the P-type calcium transport ATPases, which uses the energy of ATP molecules to transfer Ca^{2+} from cytoplasm outside the cell. In 2010 Pavlov et. al. tested another calcium ATPase—SERCA, which is structurally and functionally very similary to PMCA, for its ability to use polyP as energy source for its own activity, however, the presence of polyP did not increase the activity of SERCA (Pavlov et al. 2010).

1.3.3 Alkaline Phosphatase

Alkaline phosphatase (ALP) is an enzyme dephosphorylating different compounds. It shows its highest activity at low pH values and under physiological conditions is not very selective and can dephosphorylate different substrates including ATP, ADP, AMP and many other organic and inorganic substrates which contain orthophosphate residues (Say et al. 1991). ALP is present in all living forms starting from bacteria to mammals. In humans, it is present in all tissues. In 2001, it was shown that mammalian intestinal alkaline phosphatase has exopolyphosphatase activity and "digest" polyPs with a polymerization rate of up to 800 residues. Interestingly, ALP did not require Mg^{2+} or other divalent atoms for its activity, while other known exopolyphosphatases are highly dependent on it. However, in the presence of EDTA exopolyphosphatase activity of ALP was inhibited, that suggests some replaceable metals are crucial for the activity of the enzyme. As far as ALP contains Zn^{2+} and Mg^{2+} ions in its structure, results with EDTA look logical.

Two other papers published in 2015 and 2017 have shown that treatment of SaOS-2 cells with amorphous Ca^{2+} polyphosphate nanoparticles (Müller et al. 2015) and polyP (sodium salt) (Müller et al. 2017) increased the number of mitochondria in the cells and caused translocation of SaOS-2 cell's alkaline phosphatase to the cell surface (Müller et al. 2015). Further studies revealed that polyP treatment causes the release of matrix vesicles, which contain both ALP and adenylate kinase (AK) into the extracellular space (Müller et al. 2017). Interestingly, it was shown that pre-incubation of SaOS-2 cells with polyP increased the level of extracellular ADP and ATP. PolyP-dependent ADP increase was even higher in the presence of AK inhibitor A(5′)P5(5′)A. However, A(5′)P5(5′)A almost totally inhibited polyP-dependent ATP increase. SaOS-2 cells can release intracellular ATP via exocytosis, however, polyP-dependent ATP increase was not dependent on the presence of exocytosis inhibitors N-ethylmaleimide and brefeldin A, therefore authors concluded that the rise in ATP levels is not due to the ATP release from the cells. The authors propose that the energy stored in polyP's phosphoanhydride bonds might be used for enzymatic ATP synthesis via ALP and AK systems, where ALP is responsible for polyP dephosphorylation and AK for using the energy and orthophosphate residues from polyP degradation for nucleotide phosphorylation. AK is an enzyme, which can catalyze reaction: $ADP + ADP \leftrightarrow ATP + AMP$, in the presence of Mg^{2+}, this reaction can go in two directions, depending on the availability of the substrate. According to the proposed schemes, authors propose that the ALP degrades polyP and uses its energy for ADP synthesis from AMP (Müller et al. 2017, 2019). However, there is no evidence that ALP by dephosphorylating different substrates can phosphorylate nucleotides (AMP or ADP).

1.3.4 H-Prune

In 2008, Tammenkoski et al. demonstrated that Human Metastasis Regulator Protein (H-Prune) has polyphosphatase activity and can efficiently hydrolyze short-chain polyP (3 orthophosphates in the chain) (Tammenkoski et al. 2008). In the same research, it was shown that long-chain polyPs (more than 25 orthophosphates in the chain) not only could not be hydrolyzed by h-prune but also act as inhibitors and prevent hydrolysis of short chain polyP by h-prune (Müller et al. 2019).

1.3.5 Tartrate-Resistant Acid Phosphatase

Another enzyme, which has polyphosphatase activity, is tartrate-resistant acid phosphatase (TRAP) (Harada et al. 2013). TRAP is abundantly expressed in osteoclasts and plays an important role in bone formation. This enzyme contains iron atoms in its catalytic center, which can generate reactive oxygen species via the Fenton reaction. In this research, the authors purified TRAP from osteoblasts via coprecipitation with anti-TRAP antibodies. Experiments showed that TRAP can degrade polyP and this degradation was not dependent on ROS generation via the Fenton reaction. TRAP could degrade polyPs with different chain lengths, ranging from 40 to 750 residues; however, the speed of degradation was much higher for short and medium-chain polyPs than for long-chain polyPs (degradation of $polyP_{750}$ was almost not detectable even after 24 h of incubation). Moreover, it was found that low concentrations of polyP could inhibit phosphatase activity of TRAP, inhibiting the activity of polyP increased with the chain length of the polymer. IC_{50} for $polyP_{750}$ was ≈ 0.663 μM, for $polyP_{300} \approx 4.4$ μM, for $polyP_{40} \approx 47.6$ μM and for $polyP_{15} \approx 846$ μM. Finally, authors showed that polyP inhibited bone resorption activity of osteoclasts, based on which authors proposed that polyP might be a key molecule that regulates TRAP-mediated osteoclast bone resorption (Harada et al. 2013).

1.3.6 NAD Kinase

In 2012, Ohashi et al. for the first time identified and characterized human mitochondrial NAD kinase (Ohashi et al. 2012). NAD kinase is the sole $NADP^+$—biosynthetic enzyme known to catalyze phosphorylation of NAD^+ to yield $NADP^+$, using ATP or inorganic polyphosphate as the phosphoryl donor. This enzyme is present in all living forms, however, its activity can be different in bacteria and humans. For example, NAD kinase of some bacteria can use both ATP and polyP for NAD^+ phosphorylation, while in mammals it uses only ATP. During characterization of the newly found enzyme authors showed that it can use both ATP and polyP (hexametaphosphate and tetrapolyphosphate) for NAD^+ phosphorylation (Ohashi et al. 2012).

Acknowledgments This work was supported by the Ministry of Innovative Development of the Republic of Uzbekistan (grant number ФЗ- 20200929214). We thank Dr. T. Shiba for providing polyP standards for our research.

References

Ahn K, Kornberg A (1990) Polyphosphate kinase from *Escherichia coli*. Purification and demonstration of a phosphoenzyme intermediate. J Biol Chem 265:11734–11739

Akiyama M, Crooke E, Kornberg A (1993) An exopolyphosphatase of *Escherichia coli*. The enzyme and its *ppx* gene in a polyphosphate operon. J Biol Chem 268:633–639

Aschar-Sobbi R, Abramov AY, Diao C, Kargacin ME, Kargacin GJ, French RJ, Pavlov E (2008) High sensitivity, quantitative measurements of polyphosphate using a new DAPI-based approach. J Fluoresc 18:859–866

Baev AY, Negoda A, Abramov AY (2017) Modulation of mitochondrial ion transport by inorganic polyphosphate - essential role in mitochondrial permeability transition pore. J Bioenerg Biomembr 49:49–55

Baev AY, Angelova PR, Abramov AY (2020) Inorganic polyphosphate is produced and hydrolyzed in F_0F_1-ATP synthase of mammalian mitochondria. Biochem J 477:1515–1524

Chance B, Williams GR (1955) A simple and rapid assay of oxidative phosphorylation. Nature 175: 1120–1121

Chinopoulos C, Adam-Vizi V (2010) Mitochondria as ATP consumers in cellular pathology. Biochim Biophys Acta 1802:221–227

Das S, Lengweiler UD, Seebach D, Reusch RN (1997) Proof for a nonproteinaceous calcium-selective channel in *Escherichia coli* by total synthesis from (R)-3-hydroxybutanoic acid and inorganic polyphosphate. Proc Natl Acad Sci U S A 94:9075–9079

Elustondo PA, Nichols M, Negoda A, Thirumaran A, Zakharian E, Robertson GS, Pavlov EV (2016) Mitochondrial permeability transition pore induction is linked to formation of the complex of ATPase C-subunit, polyhydroxybutyrate and inorganic polyphosphate. Cell Death Discov 2:16070

Harada K, Itoh H, Kawazoe Y, Miyazaki S, Doi K, Kubo T, Akagawa Y, Shiba T (2013) Polyphosphate-mediated inhibition of tartrate-resistant acid phosphatase and suppression of bone resorption of osteoclasts. PLoS One 8:e78612

Kim D, Cavanaugh EJ (2007) Requirement of a soluble intracellular factor for activation of transient receptor potential A1 by pungent chemicals: role of inorganic polyphosphates. J Neurosci 27:6500–6509

Kornberg A (1999) Inorganic polyphosphate: a molecule of many functions. Prog Mol Subcell Biol 23:1–18

Kumble KD, Kornberg A (1996) Endopolyphosphatases for long chain inorganic polyphosphate in yeast and mammals. J Biol Chem 271:27146–27151

Mitchell P, Moyle J (1968) Proton translocation coupled to ATP hydrolysis in rat liver mitochondria. Eur J Biochem 4:530–539

Müller WEG, Tolba E, Feng QL, Schröder HC, Markl JS, Kokkinopoulou M, Wang XH (2015) Amorphous Ca^{2+} polyphosphate nanoparticles regulate the ATP level in bone-like SaOS-2 cells. J Cell Sci 128:2202–2207

Müller WEG, Wang SF, Neufurth M, Kokkinopoulou M, Feng QL, Schröder HC, Wang XH (2017) Polyphosphate as a donor of high-energy phosphate for the synthesis of ADP and ATP. J Cell Sci 130:2747–2756

Müller WEG, Schröder HC, Wang XH (2019) Inorganic polyphosphates as storage for and generator of metabolic energy in the extracellular matrix. Chem Rev 119:12337–12374

Negoda A, Xian M, Reusch RN (2007) Insight into the selectivity and gating functions of *Streptomyces lividans* KcsA. Proc Natl Acad Sci U S A 104:4342–4346

Negoda A, Negoda E, Xian M, Reusch RN (2009) Role of polyphosphate in regulation of the Streptomyces lividans KcsA channel. Biochim Biophys Acta 1788:608–614

O'Boyle NM, Banck M, James CA, Morley C, Vandermeersch T, Hutchison GR (2011) Open Babel: an open chemical toolbox. J Cheminform 3:33

Ohashi K, Kawai S, Murata K (2012) Identification and characterization of a human mitochondrial NAD kinase. Nat Commun 3:1248

Pavlov E, Zakharian E, Bladen C, Diao CT, Grimbly C, Reusch RN, French RJ (2005) A large, voltage-dependent channel, isolated from mitochondria by water-free chloroform extraction. Biophys J 88:2614–2625

Pavlov E, Aschar-Sobbi R, Campanella M, Turner RJ, Gomez-Garcia MR, Abramov AY (2010) Inorganic polyphosphate and energy metabolism in mammalian cells. J Biol Chem 285:9420–9428

Pettersen EF, Goddard TD, Huang CC, Couch GS, Greenblatt DM, Meng EC, Ferrin TE (2004) UCSF Chimera—a visualization system for exploratory research and analysis. J Comput Chem 25:1605–1612

Reusch RN, Huang R, Bramble LL (1995) Poly-3-hydroxybutyrate/polyphosphate complexes form voltage-activated Ca^{2+} channels in the plasma membranes of *Escherichia coli*. Biophys J 69: 754–766

Reusch RN, Huang R, Kosk-Kosicka D (1997) Novel components and enzymatic activities of the human erythrocyte plasma membrane calcium pump. FEBS Lett 412:592–596

Roewe J, Stavrides G, Strueve M, Sharma A, Marini F, Mann A, Smith SA, Kaya Z, Strobl B, Mueller M, Reinhardt C, Morrissey JH, Bosmann M (2020) Bacterial polyphosphates interfere with the innate host defense to infection. Nat Commun 11:4035

Say JC, Ciuffi K, Furriel RP, Ciancaglini P, Leone FA (1991) Alkaline phosphatase from rat osseous plates: purification and biochemical characterization of a soluble form. Biochim Biophys Acta 1074:256–262

Solesio ME, Elustondo PA, Zakharian E, Pavlov EV (2016) Inorganic polyphosphate (polyP) as an activator and structural component of the mitochondrial permeability transition pore. Biochem Soc Trans 44:7–12

Spikes TE, Montgomery MG, Walker JE (2020) Structure of the dimeric ATP synthase from bovine mitochondria. Proc Natl Acad Sci U S A 117:23519–23526

Tammenkoski M, Koivula K, Cusanelli E, Zollo M, Steegborn C, Baykov AA, Lahti R (2008) Human metastasis regulator protein H-prune is a short-chain exopolyphosphatase. Biochemistry 47:9707–9713

Trott O, Olson AJ (2010) AutoDock Vina: improving the speed and accuracy of docking with a new scoring function, efficient optimization, and multithreading. J Comput Chem 31:455–461

Wurst H, Kornberg A (1994) A soluble exopolyphosphatase of *Saccharomyces cerevisiae*. Purification and characterization. J Biol Chem 269:10996–11001

Zakharian E, Thyagarajan B, French RJ, Pavlov E, Rohacs T (2009) Inorganic polyphosphate modulates TRPM8 channels. PLoS One 4:e5404

Zorova LD, Popkov VA, Plotnikov EY, Silachev DN, Pevzner IB, Jankauskas SS, Babenko VA, Zorov SD, Balakireva AV, Juhaszova M, Sollott SJ, Zorov DB (2018) Mitochondrial membrane potential. Anal Biochem 552:50–59

Chapter 2
Inorganic Polyphosphate in Mitochondrial Energy Metabolism and Pathology

Maria A. Neginskaya and Evgeny V. Pavlov

Abstract In this chapter, the current understanding of the potential roles played by polyphosphate in mitochondrial function with a specific focus on energy metabolism and mitochondrial pathologies caused by stress is summarized. Here we will discuss details of the possible ion transporting mechanisms of mitochondria that might involve polyP and their role in mitochondrial physiology and pathology are discussed.

Keywords Mitochondrial function · Polyphosphate kinase · PolyP hydrolyzing enzyme · Calcium · Permeability transition pore · Oxidative phosphorylation · ATP · Polyhydroxybutyrate

2.1 Introduction

Mitochondria are an intracellular organelles that are present in nearly all eukaryotic cells. Mitochondria play several key functions, including participation in energy production through the generation of ATP, calcium, and ROS signaling and regulation of cell death. Mitochondrial ability to produce ATP is linked to the function of an oxidative phosphorylation system (OXPHOS). This system uses respiratory chain enzymes and converts the energy stored in NADH and FADH into the energy of electrochemical gradient across the mitochondrial inner membrane. This energy is used to generate ATP by the ATP synthase complex. The nature of the chemical phosphate bonds in polyP and ATP is very similar, and it has been demonstrated that in bacteria, these two phosphate polymers can be converted into each other using polyphosphate kinase (PPK), a specific enzyme without any energy source (Ahn and Kornberg 1990). Here we will review the current knowledge of the possible relationship between polyP and ATP metabolism in mitochondria and discuss possible similarities and differences. Another important aspect of mitochondrial function is

M. A. Neginskaya · E. V. Pavlov (✉)
Department of Molecular Pathobiology, New York University, New York, NY, USA
e-mail: mn2452@nyu.edu; ep37@nyu.edu

W. E. G. Müller et al. (eds.), *Inorganic Polyphosphates*, Progress in Molecular and Subcellular Biology 61, https://doi.org/10.1007/978-3-031-01237-2_2

its central role as a hub of cellular calcium signaling (Duchen 2000). Mitochondria are capable of accumulating calcium in response to either physiological or pathological stimuli. Levels of mitochondrial free calcium are closely linked to the rates of mitochondrial metabolic activity (Glancy and Balaban 2012). The relationship between total and free mitochondrial calcium is tightly regulated by different phosphate species (Chalmers and Nicholls 2003). PolyP as a highly charged negative polymer has been demonstrated to form complexes with calcium and regulate calcium–phosphate interactions in vitro and in microorganisms (Omelon and Grynpas 2008). We will discuss if some similar mechanisms might be present in mitochondria as well. PolyP has been implicated as a regulator and structural component of several ion channels (Das et al. 1997; Pavlov et al. 2005; Zakharian and Reusch 2004, 2007; Zakharian et al. 2009). Here we will discuss details of the possible ion transporting mechanisms that might involve polyP and their role in mitochondrial physiology and pathology.

Throughout the history of polyP studies in mammalian mitochondria and in higher eukaryotes in general, the methodological issues linked to its accurate assay have been the major challenge. As such, the experimental data addressing the role of mitochondrial polyP in the specific mitochondrial function need to be carefully considered in the context of a specific assay or experimental method. Indeed, polyP is a negatively charged polymer capable of interacting and forming complexes with many molecules. This makes it very difficult to isolate polyP from the biological samples for quantification as large amounts could be left in the sample in bound form. Further, the lack of a specific group makes it a challenge to develop a probe that will be selective specifically to polyP and not interact with other molecules (e.g., DNA and RNA). Taking this into consideration, it is important to carefully review and estimate the possibility that some nonspecific interaction might contribute to the polyP signal and interpret data accordingly. Whenever possible, we will try to briefly describe an experimental design that was the basis for specific conclusions regarding polyP function.

2.2 Early Studies of Mitochondrial PolyP and Methods for Its Assay

The multiple roles of polyP in the living nature have been most extensively investigated in bacterial organisms. It has been established that in bacteria, polyP can serve as energy storage as well as can be involved in the regulation of divalent ions buffering, the role that can be easily envisioned for the mitochondria (Kornberg et al. 1999). Based on the results of polyP studies in bacteria, it has been proposed that like in simple organisms polyP might be contributing importantly to the function of mitochondria through similar mechanisms. However, one of the critical arguments that polyP mechanisms in mitochondria might be very different is linked to the observation that levels of polyP in mitochondria are orders of magnitude lower compared to bacteria. In the very early polyP studies, the only report that directly addressed the possible polyP metabolism in mammals involved the use of isolated rat liver mitochondria. In this work, mitochondria were treated with various energy substrates and measurements of polyP following its isolation from these samples (Lynn and Brown 1963). However, due to the low levels of the polyP in these samples, it was argued that assays were not specific enough and thus did not reflect the true levels of this polymer in mitochondria (Harold 1966). Overall, it was generally concluded that polyP might not even be present in eukaryotes where evolutionary it was substituted by other types of polyphosphates such as ATP (Harold 1966). Some roles of polyP might also be shared with inositol pyrophosphates which were summarized in a recent review (Azevedo and Saiardi 2017). The more definitive proof that polyP is indeed present in higher eukaryotes and specifically in mammals was provided later following the development of the specific enzymatic assay that was based on the reaction catalyzed by polyP kinase (PPK) from bacteria. PPK is an enzyme that can either produce polyP from ATP or produce ATP using polyP as a substrate. In experiments with mammalian mitochondrial samples, it was demonstrated that the addition of PPK to the extract leads to the production of ATP (Kumble and Kornberg 1995). This was unambiguous proof of the presence of polyP in mammals. Another important conclusion from this chapter was that mammalian cultured cells have high turnover rates of polyP, confirming that polyP is not only present in these samples, but that it can also be actively produced and consumed. These two observations established the foundation for future studies of the possible polyP metabolic pathways and their roles in mammalian cells and tissues. There are two important outstanding questions regarding mitochondrial polyP that are of critical importance following the results of this seminal paper. Firstly, this study assessed the total amounts of polyP and did not discriminate between free and bound polymers. Like other highly charged ions, it is very likely that the bulk of polyP is present in the complexed form, and thus assay of the polyP does not allow to conclude which fraction of this polymer is bioavailable to perform such functions as an energy substrate for the enzyme or as divalent ions buffer. As a result, the changes in total polyP might not reflect the changes in the specific polyP pools involved in specific biological roles. Further, mitochondrial polyP was measured in isolated mitochondria lacking substrates. Since these mitochondria were not in the active metabolic state, the levels of polyP might not reflect the true levels of polyP that would be found in the fully functional organelles. In an effort to dissect specific roles associated with specific forms of mitochondrial polyP, the follow-up studies tried to look specifically at the changes in the free polymer in living cells or in isolated mitochondria as a function of their metabolic state. This type of assay can be performed using fluorescent probes that are selective to the polymer. In this case, polyP changes can be detected in real-time in living organelles (Angelova et al. 2014; Aschar-Sobbi et al. 2008; Pavlov et al. 2010). However, due to the chemical nature of polyP, such fluorescent probes are relatively weakly selective and, in many cases, can interact with other polyphosphates (RNA, DNA) or charged macromolecules. As a result, data interpretation of these experiments is not always straightforward. Other types of assays involved studies of the effects of the synthetic polyP added to the living cells (Holmstrom et al. 2013; Stotz et al. 2014). These assays allow us to conclusively determine what effects polyP can induce; however, they cannot clarify whether similar effects could be expected from the endogenous polymer. It is also not entirely clear how specific effects of polyP are and if they might be caused simply due to the presentation of the highly-dense negative charge. In addition, the effects of polyP could be studied indirectly by the effects of the expression of the polyP hydrolyzing enzyme (PPX) (Akiyama et al. 1993). PPX selectively hydrolyzes polyP and provides a model that can compare how the function of the cell and mitochondria are different if polyP is not present (Abramov et al. 2007; Pavlov et al. 2010; Seidlmayer et al. 2012, 2015). In the following sections, we will discuss in more detail the results of the experiments obtained using these techniques and their limitations that should be taken into consideration when making conclusions about polyP activities and metabolism.

2.3 PolyP as an Energy Molecule for the Mitochondrial Function

Mitochondrial energy metabolic pathway results in the generation of the ATP molecule through the process of OXPHOS that involves the generation of the membrane potential by the respiratory chain, the energy of which is used by the ATP synthase enzyme to produce ATP from the ADP. In bacteria, ATP can be utilized to generate polyP (Rao et al. 2009). Considering that this reaction is reversible, polyP can be used as an energy storage polymer that, if needed, can be converted back into ATP. In bacteria storage, polyP is generated by the PPK enzyme. The existence of polyP in mitochondria suggests the possibility that it could potentially be used as an energy source. However, the synthesis and utilization pathway most likely is very different. Mammalian mitochondria do not have a PPK enzyme homolog. This is consistent with the observation that bacteria PPK produces polyP of a very high (up to 1000 units) length, which has not been found in mammalian mitochondria. The relationship between mitochondrial energy metabolism and levels of polyP has been first investigated in early studies that used rat liver mitochondria (Lynn and Brown 1963). In these experiments, isolated mitochondria were incubated with succinate as a substrate. It was reported that stimulation of mitochondria with succinate resulted in an increase in polyP production. Importantly, the addition of ADP resulted in a decrease in the amounts of recovered polyP. The authors concluded that in the absence of the source of ADP, mitochondria could produce polyP presumably as an energy storage compound. While these results were very promising, it should be mentioned that the finding of this chapter was later subject of the discussion of the possibility that extracted compounds might not have represented polyP, although no follow-up study was performed. Another attempt to link mitochondrial bioenergetics state with polyP production was performed more recently. Similar to the earlier works in this study, isolated rat liver mitochondria were stimulated by various substrates, and changes in polyP levels were monitored in real time using a DAPI fluorescence assay that was optimized for polyP detection (Pavlov et al. 2010). This assay method had very important differences compared to the previously used assays that relied on the extraction of the total polyP pool from the samples. The advantage of DAPI is that it interacts with a free polymer which makes it particularly useful for an estimate of the bioavailable pool of polyP rather than the total polyP that includes the bulk of the polymer that is bound to proteins, DNA and RNA and not necessarily involved in energy metabolism. With the use of this approach, it was found that stimulation of the mitochondria leads to the increase of DAPI fluorescence, presumably reflecting an increase in polyP production. Importantly, these changes were sensitive to the inhibitors of the respiratory chain and ATPase suggesting the direct link between energy production and polyP generation. This study did not use a biochemical assay to quantify the changes in the amounts of polyP in each tested condition. However, it confirmed biochemically that polyP amounts change with changes in the metabolic state of the mitochondria caused by the mitochondrial depolarization. In this particular case, DAPI fluorescence assay correlated with biochemical measurements. This biochemical assay also demonstrated that not only the amount, but the chain length of polyP pool undergoes changes with polyP length becoming increasingly shorter when OXPHOS is uncoupled by the addition of ionophore FCCP. The finding that stimulation of mitochondrial metabolism leads to production of polyP was also confirmed at the level of intact cells. In these experiments, DAPI fluorescence decreased in response to inhibition of glycolysis, respiratory chain, mitochondrial membrane depolarization, and inhibition of ATPase but increased upon the addition of energy substrates. It should be noted, however, that although DAPI can be optimized to label polyP preferentially, it is still only moderately selective, particularly when interacting with RNA. Further, when DAPI is dehydrated, it can demonstrate a similar redshift as in the case of DAPI-polyP interaction. Thus, data obtained using DAPI probe should be interpreted with care.

It is also noteworthy that more recent experiments demonstrate that polyP can directly regulate activity of the mitochondrial ATPase (Baev et al. 2020). In these experiments, addition of various amounts of polyP induced changes in mitochondrial function. The authors propose that these experimental data are consistent with the possibility that the ATPase enzyme can be capable of metabolizing polyP. It will be interesting to test by the direct assay the proposed hypothesis that ATPase might indeed play such a role.

Overall, experimental data currently available support the idea of the correlation between the rates of polyP production and consumption and mitochondrial energy metabolism. At the same time, the question regarding the function of specific enzymes directly involved in polyP turnover remains open. In microorganisms, polyP metabolism involves specialized enzymes (Kornberg et al. 1999). PolyP can be produced by PPK that uses ATP as a substrate. Interestingly, the enzyme is reversible and can also consume polyP by using it as a substrate for ATP production. PolyP can also be hydrolyzed by exo- (PPX) and endo- (PPN) polyphosphatases. PPX progressively cleaves orthophosphate groups from the end of the polymer while

PPN splits the polyP chain in the middle. It is very likely that more polyP metabolizing enzymes exist. For example, PPK knockout in *E. coli* leads to almost complete elimination of long-chain polyP. At the same time, levels of the medium-chain polyP (polymer made of ~100 orthophosphate units) remained virtually unchanged, suggesting the existence of separate pathways responsible for the production of specific pools of polyP (Castuma et al. 1995). The identification of specific mitochondrial polyP metabolizing enzymes remains to be the key task that will help significantly advance our understanding of the role of this polymer in mitochondria. Further, eukaryotic polyP has been recently proposed as an energy source for the extracellular matrix (Müller et al. 2017, 2019). It will be very interesting to explore if mitochondrial polyP is linked to this extracellular pool.

2.4 PolyP Role as a Signaling Molecule and Regulator of the Mitochondrial Calcium Buffering Capacity

Another prominent role of polyP in mitochondrial function that is not linked directly to bioenergetics is its involvement in the regulation of calcium hemostasis. Calcium plays a very important role in the regulation of mitochondrial function in both normal physiology and pathology. Mitochondria can accumulate calcium inside the matrix by electrogenic uptake (Bernardi 1999; Duchen 2000; Elustondo et al. 2017; Rizzuto et al. 2000). Levels of free calcium inside mitochondria are regulated by the rates of uptake and release and by the mitochondrial calcium buffering by phosphate (Chalmers and Nicholls 2003; Nicholls and Chalmers 2004). The importance of phosphate is clear from the experiments when orthophosphate was removed, which resulted in the complete loss of the regulation of the mitochondrial free calcium (Chalmers and Nicholls 2003). Calcium-phosphate buffering is also of great importance during the excessive calcium overload in pathology. In these conditions, despite continuous calcium uptake, the concentration of the free calcium remains relatively constant while calcium-phosphate precipitates increase dramatically. This rise in calcium-phosphate continues until activation of the mitochondrial permeability transition pore causing mitochondrial depolarization and loss of function. Previous studies on tissue calcification demonstrated that polyP could play an important role in the regulation of calcium-phosphate precipitation (Omelon and Grynpas 2008). Specifically, it has been proposed that the presence of polyP leads to inhibition of precipitation. Recent data suggest that polyP might be involved in the regulation of the mitochondrial free calcium as well, presumably through the regulation of the calcium–phosphate interactions (Solesio et al. 2016, 2020).

The importance of polyP as a signal polymer—the regulator of the energy metabolism became more evident with the recent study that investigated the relationship between rates of OXPHOS and levels of mitochondrial polyP (Solesio et al. 2021). In this study, using metabolomics and functional assays, it was demonstrated how depletion of polyP affects cell metabolism in general. It was established that

HEK293 cells that stably express the PPX enzyme have significantly altered metabolism. Specifically, the lack of mitochondrial polyP caused the metabolic shift from OXPHOS to glycolysis. This was confirmed both by measuring rates of oxygen consumption and changes in extracellular pH (as a measure of glycolysis) and by measuring levels of metabolites—markers of the specific pathway. Importantly, at the level of isolated mitochondria, the oxygen consumption rates were similar in both wild type and polyP(-) mitochondria. This suggests that the effects of polyP were determined not by its direct regulation of the mitochondrial function or through its involvement in ATP/ADP turnover but rather through other mechanisms. Currently, the exact pathways that are responsible for such a switch are not known. Several possibilities can be considered, which would involve polyP redistribution from mitochondria and stimulation of glycolysis. Another option might be linked to the above-mentioned role of polyP in the regulation of the levels of mitochondrial free calcium. Increased levels of free calcium are known to be involved in the stimulation of the OXPHOS. It is possible that loss in the mitochondrial polyP can cause a loss in the ability of the cellular calcium signaling system to stimulate the OXPHOS and, as a result, to the increased role of glycolysis in the cellular ATP supply.

2.5 PolyP, Membrane Ion Transport, and the Permeability Transition Pore (PTP)

One of the least studied, and most intriguing roles that polyP might play in mitochondrial function is its potential contribution to the transmembrane ion transport. The role of polyP as an ion transporting molecule has first been proposed for bacteria. In her seminal work that was published in 1988, Dr. Reusch has reported that significant amounts of polyP present in *E. coli* can be found in the complex with polyhydroxy butyrate (PHB) and calcium (polyP/Ca^{2+}/PHB complex) (Reusch et al. 1995; Reusch and Sadoff 1988). She reported that this complex purified from bacteria could form calcium-selective channels when reconstituted into model planar lipid bilayers. Later channel-forming activity was confirmed by experiments with a completely synthetic complex (Das et al. 1997). Taking into account that amounts of the complex increased dramatically in the competent bacteria it was proposed that it might play a role in DNA transport across bacterial membranes (Huang and Reusch 1995). Interestingly, the same complex was found in many different organisms, including mammalian mitochondria (Reusch 1989). As this channel demonstrated calcium selectivity, it was initially hypothesized that it could play a role in mitochondrial calcium uptake. However, later experiments discovered that purified mitochondrial complex could form large pores with properties resembling properties of the Mitochondrial Permeability Transition Pore (PTP) (Pavlov et al. 2005). PTP is a large nonselective channel (pore) that forms in the mitochondrial inner membrane in response to excessive calcium accumulation (Zoratti and Szabo 1995). The ion conducting behavior of the PTP is highly variable (Neginskaya et al. 2021), which suggests that it can be formed by several molecular mechanisms. It is currently believed that primary PTP pathways can involve either ATP synthase (Alavian et al. 2014; Bonora et al. 2013; Giorgio et al. 2013; Neginskaya et al. 2019) or Adenine Nucleotide Translocase (ANT) (Brustovetsky and Klingenberg 1996; Karch et al. 2019). Presence of polyP/Ca^{2+}/PHB in calcium-treated mitochondria raises an intriguing possibility that this complex might provide an alternative PTP pathway or participate in its formation through the interaction with ATP synthase or ANT. In agreement with this hypothesis, later work confirmed that depletion of the mitochondrial polyP leads to the inhibition of the PTP as well as prevents calcium-induced death of cultured neurons and HepG2 cells (Abramov et al. 2007). Similar inhibition has also been found in primary cultures of cardiomyocytes (Seidlmayer et al. 2012). Interestingly experiments with cardiomyocytes demonstrated that PTP could not be inhibited when activated not by calcium but by the reactive oxygen species (ROS) during oxidative stress (Seidlmayer et al. 2015). This suggests that mechanisms linking levels of polyP with mitochondrial resistance against stress might be very different depending on the specific stress conditions. One of the possible explanations for why polyP is important in protection against ROS-induced PTP is its recently described role as a chaperone that can prevent protein misfolding (Cremers et al. 2016; Gray et al. 2014). Although this pathway has not been experimentally verified for the mitochondria, in vitro experiments support the possibility of such a role for polyP. More recent studies that revisited the possible role of polyP in calcium-induced PTP indicate that similar to the competent bacteria amounts of polyP/Ca^{2+}/PHB complex in mitochondria are increased in response to the conditions of increased calcium accumulation—a condition that favors PTP opening (Elustondo et al. 2016). Interestingly, this increase coincides with the increase in the fraction of the c subunit of the ATP synthase that co-purifies with polyP and PHB fractions. This phenomenon was seen in both isolated mitochondria and in mouse brain tissues following stroke. Taking into account the c subunit is one of the central membrane proteins involved in the formation of the PTP it is tantalizing to hypothesize that during stress, formation of the PTP might involve interactions between the c subunit and polyP. With this respect it is worth noting that recent studies uncovered that the c subunit of the ATP synthase is an amyloid peptide. This further supports the idea that polyP participation in PTP might involve specific interactions with misfolded copies of the peptide rather with the peptide in its native form (Amodeo et al. 2021; Amodeo and Pavlov 2021). This interpretation is favored by the observation from the same group that each mitochondrion has a very small number of PTP channels despite a large number of fully folded copies of the ATP synthase (Neginskaya et al. 2020).

2.6 Future Directions

While the investigation of polyP in higher eukaryotes still remains a relatively small field of research, it draws an increasing number of scientists from various areas. With this respect, investigation of polyP as a metabolite, regulator, and structural component of mitochondria is an important aspect with high relevance to many areas of life sciences. Perhaps the most critical question that needs to be answered is about the enzymatic process responsible for polyP production. Another aspect of research that needs further development is the improvement of polyP detection methods. Indeed, many of the statements regarding the roles of polyP in mitochondria would remain somewhat speculative until they can be verified by selective and sensitive experimental assays. Finally, the role of polyP in calcium handling could be further explored with the utilization of such experimental methodologies as Cryo-EM tomography which would allow to monitor and quantify polyP-calcium precipitates that cannot be detected by fluorescent probes.

References

Abramov AY, Fraley C, Diao CT, Winkfein R, Colicos MA, Duchen MR, French RJ, Pavlov E (2007) Targeted polyphosphatase expression alters mitochondrial metabolism and inhibits calcium-dependent cell death. Proc Natl Acad Sci USA 104:18091–18096

Ahn K, Kornberg A (1990) Polyphosphate kinase from *Escherichia coli*. Purification and demonstration of a phosphoenzyme intermediate. J Biol Chem 265:11734–11739

Akiyama M, Crooke E, Kornberg A (1993) An exopolyphosphatase of *Escherichia coli*. The enzyme and its *ppx* gene in a polyphosphate operon. J Biol Chem 268:633–639

Alavian KN, Beutner G, Lazrove E, Sacchetti S, Park HA, Licznerski P, Li H, Nabili P, Hockensmith K, Graham M et al (2014) An uncoupling channel within the c-subunit ring of the F1FO ATP synthase is the mitochondrial permeability transition pore. Proc Natl Acad Sci USA 111:10580–10585

Amodeo GF, Pavlov EV (2021) Amyloid beta, alpha-synuclein and the c subunit of the ATP synthase: can these peptides reveal an amyloidogenic pathway of the permeability transition pore? Biochim Biophys Acta Biomembr 1863:183531

Amodeo GF, Lee BY, Krilyuk N, Filice CT, Valyuk D, Otzen DE, Noskov S, Leonenko Z, Pavlov EV (2021) C subunit of the ATP synthase is an amyloidogenic calcium dependent channel-forming peptide with possible implications in mitochondrial permeability transition. Sci Rep 11:8744

Angelova PR, Agrawalla BK, Elustondo PA, Gordon J, Shiba T, Abramov AY, Chang YT, Pavlov EV (2014) *In situ* investigation of mammalian inorganic polyphosphate localization using novel selective fluorescent probes JC-D7 and JC-D8. ACS Chem Biol 9:2101–2110

Aschar-Sobbi R, Abramov AY, Diao C, Kargacin ME, Kargacin GJ, French RJ, Pavlov E (2008) High sensitivity, quantitative measurements of polyphosphate using a new DAPI-based approach. J Fluoresc

Azevedo C, Saiardi A (2017) Eukaryotic phosphate homeostasis: the inositol pyrophosphate perspective. Trends Biochem Sci 42:219–231

Baev AY, Angelova PR, Abramov AY (2020) Inorganic polyphosphate is produced and hydrolyzed in F_0F_1-ATP synthase of mammalian mitochondria. Biochem J 477:1515–1524

Bernardi P (1999) Mitochondrial transport of cations: channels, exchangers, and permeability transition. Physiol Rev 79:1127–1155

Bonora M, Bononi A, De ME, Giorgi C, Lebiedzinska M, Marchi S, Patergnani S, Rimessi A, Suski JM, Wojtala A et al (2013) Role of the c subunit of the F_O ATP synthase in mitochondrial permeability transition. Cell Cycle 12:674–683

Brustovetsky N, Klingenberg M (1996) Mitochondrial ADP/ATP carrier can be reversibly converted into a large channel by Ca^{2+}. Biochemistry 35:8483–8488

Castuma CE, Huang R, Kornberg A, Reusch RN (1995) Inorganic polyphosphates in the acquisition of competence in *Escherichia coli*. J Biol Chem 270:12980–12983

Chalmers S, Nicholls DG (2003) The relationship between free and total calcium concentrations in the matrix of liver and brain mitochondria. J Biol Chem 278:19062–19070

Cremers CM, Knoefler D, Gates S, Martin N, Dahl JU, Lempart J, Xie L, Chapman MR, Galvan V, Southworth DR et al (2016) Polyphosphate: a conserved modifier of amyloidogenic processes. Mol Cell 63:768–780

Das S, Lengweiler UD, Seebach D, Reusch RN (1997) Proof for a nonproteinaceous calcium-selective channel in *Escherichia coli* by total synthesis from (R)-3-hydroxybutanoic acid and inorganic polyphosphate. Proc Natl Acad Sci USA 94:9075–9079

Duchen MR (2000) Mitochondria and Ca^{2+} in cell physiology and pathophysiology. Cell Calcium 28:339–348

Elustondo PA, Nichols M, Negoda A, Thirumaran A, Zakharian E, Robertson GS, Pavlov EV (2016) Mitochondrial permeability transition pore induction is linked to formation of the complex of ATPase C-subunit, polyhydroxybutyrate and inorganic polyphosphate. Cell Death Discov 2:16070

Elustondo PA, Nichols M, Robertson GS, Pavlov EV (2017) Mitochondrial Ca^{2+} uptake pathways. J Bioenerg Biomembr 49:113–119

Giorgio V, von SS, Antoniel M, Fabbro A, Fogolari F, Forte M, Glick GD, Petronilli V, Zoratti M, Szabo I, et al. (2013) Dimers of mitochondrial ATP synthase form the permeability transition pore. Proc Natl Acad Sci USA 110:5887–5892

Glancy B, Balaban RS (2012) Role of mitochondrial Ca^{2+} in the regulation of cellular energetics. Biochemistry 51:2959–2973

Gray MJ, Wholey WY, Wagner NO, Cremers CM, Mueller-Schickert A, Hock NT, Krieger AG, Smith EM, Bender RA, Bardwell JC et al (2014) Polyphosphate is a primordial chaperone. Mol Cell 53:689–699

Harold FM (1966) Inorganic polyphosphates in biology: structure, metabolism, and function. Bacteriol Rev 30:772–794

Holmstrom KM, Marina N, Baev AY, Wood NW, Gourine AV, Abramov AY (2013) Signalling properties of inorganic polyphosphate in the mammalian brain. Nat Commun 4:1362

Huang R, Reusch RN (1995) Genetic competence in *Escherichia coli* requires poly-beta-hydroxybutyrate/calcium polyphosphate membrane complexes and certain divalent cations. J Bacteriol 177:486–490

Karch J, Bround MJ, Khalil H, Sargent MA, Latchman N, Terada N, Peixoto PM, Molkentin JD (2019) Inhibition of mitochondrial permeability transition by deletion of the ANT family and CypD. Sci Adv 5(eaaw4597)

Kornberg A, Rao NN, Ault-Riche D (1999) Inorganic polyphosphate: a molecule of many functions. Annu Rev Biochem 68:89–125

Kumble KD, Kornberg A (1995) Inorganic polyphosphate in mammalian cells and tissues. J Biol Chem 270:5818–5822

Lynn WS, Brown RH (1963) Synthesis of polyphosphate by rat liver mitochondria. Biochem Biophys Res Commun 11:367–371

Müller WEG, Wang SF, Neufurth M, Kokkinopoulou M, Feng QL, Schröder HC, Wang XH (2017) Polyphosphate as a donor of high-energy phosphate for the synthesis of ADP and ATP. J Cell Sci 130:2747–2756

Müller WEG, Schröder HC, Wang XH (2019) Inorganic polyphosphates as storage for and generator of metabolic energy in the extracellular matrix. Chem Rev 119:12337–12374

Neginskaya MA, Solesio ME, Berezhnaya EV, Amodeo GF, Mnatsakanyan N, Jonas EA, Pavlov EV (2019) ATP synthase C-subunit-deficient mitochondria have a small cyclosporine A-sensitive channel, but lack the permeability transition pore. Cell Rep 26(11–17):e12

Neginskaya MA, Strubbe JO, Amodeo GF, West BA, Yakar S, Bazil JN, Pavlov EV (2020) The very low number of calcium-induced permeability transition pores in the single mitochondrion. J Gen Physiol 152

Neginskaya MA, Pavlov EV, Sheu SS (2021) Electrophysiological properties of the mitochondrial permeability transition pores: channel diversity and disease implication. Biochim Biophys Acta Bioenerg 1862:148357

Nicholls DG, Chalmers S (2004) The integration of mitochondrial calcium transport and storage. J Bioenerg Biomembr 36:277–281

Omelon SJ, Grynpas MD (2008) Relationships between polyphosphate chemistry, biochemistry and apatite biomineralization. Chem Rev 108:4694–4715

Pavlov E, Zakharian E, Bladen C, Diao CT, Grimbly C, Reusch RN, French RJ (2005) A large, voltage-dependent channel, isolated from mitochondria by water-free chloroform extraction. Biophys J 88:2614–2625

Pavlov E, Aschar-Sobbi R, Campanella M, Turner RJ, Gomez-Garcia MR, Abramov AY (2010) Inorganic polyphosphate and energy metabolism in mammalian cells. J Biol Chem 285(13): 9420–9428

Rao NN, Gomez-Garcia MR, Kornberg A (2009) Inorganic polyphosphate: essential for growth and survival. Annu Rev Biochem 78:605–647

Reusch RN (1989) Poly-beta-hydroxybutyrate/calcium polyphosphate complexes in eukaryotic membranes. Proc Soc Exp Biol Med 191:377–381

Reusch RN, Sadoff HL (1988) Putative structure and functions of a poly-beta-hydroxybutyrate/ calcium polyphosphate channel in bacterial plasma membranes. Proc Natl Acad Sci U S A 85: 4176–4180

Reusch RN, Huang R, Bramble LL (1995) Poly-3-hydroxybutyrate/polyphosphate complexes form voltage-activated Ca^{2+} channels in the plasma membranes of Escherichia coli. Biophys J 69: 754–766

Rizzuto R, Bernardi P, Pozzan T (2000) Mitochondria as all-round players of the calcium game. J Physiol 529(Pt 1):37–47

Seidlmayer LK, Gomez-Garcia MR, Blatter LA, Pavlov E, Dedkova EN (2012) Inorganic polyphosphate is a potent activator of the mitochondrial permeability transition pore in cardiac myocytes. J Gen Physiol 139:321–331

Seidlmayer LK, Juettner VV, Kettlewell S, Pavlov EV, Blatter LA, Dedkova EN (2015) Distinct mPTP activation mechanisms in ischaemia-reperfusion: contributions of Ca^{2+}, ROS, pH, and inorganic polyphosphate. Cardiovasc Res 106:237–248

Solesio ME, Demirkhanyan L, Zakharian E, Pavlov EV (2016) Contribution of inorganic polyphosphate towards regulation of mitochondrial free calcium. Biochim Biophys Acta 1860:1317–1325

Solesio ME, Garcia Del Molino LC, Elustondo PA, Diao C, Chang JC, Pavlov EV (2020) Inorganic polyphosphate is required for sustained free mitochondrial calcium elevation, following calcium uptake. Cell Calcium 86:102127

Solesio ME, Xie L, McIntyre B, Ellenberger M, Mitaishvili E, Bhadra-Lobo S, Bettcher LF, Bazil JN, Raftery D, Jakob U et al (2021) Depletion of mitochondrial inorganic polyphosphate (polyP) in mammalian cells causes metabolic shift from oxidative phosphorylation to glycolysis. Biochem J 478:1631–1646

Stotz SC, Scott LO, Drummond-Main C, Avchalumov Y, Girotto F, Davidsen J, Gomez-Garcia MR, Rho JM, Pavlov EV, Colicos MA (2014) Inorganic polyphosphate regulates neuronal excitability through modulation of voltage-gated channels. Mol Brain 7:42

Zakharian E, Reusch RN (2004) Functional evidence for a supramolecular structure for the *Streptomyces lividans* potassium channel KcsA. Biochem Biophys Res Commun 322:1059–1065

Zakharian E, Reusch RN (2007) *Haemophilus influenzae* outer membrane protein P5 is associated with inorganic polyphosphate and polyhydroxybutyrate. Biophys J 92:588–593

Zakharian E, Thyagarajan B, French RJ, Pavlov E, Rohacs T (2009) Inorganic polyphosphate modulates TRPM8 channels. PLoS One 4:e5404

Zoratti M, Szabo I (1995) The mitochondrial permeability transition. Biochim Biophys Acta 1241: 139–176

Chapter 3
Inorganic Polyphosphate, Mitochondria, and Neurodegeneration

Pedro Urquiza and Maria E. Solesio

Abstract With an aging population, the presence of aging-associated pathologies is expected to increase within the next decades. Regrettably, we still do not have any valid pharmacological or non-pharmacological tools to prevent, revert, or cure these pathologies. The absence of therapeutical approaches against aging-associated pathologies can be at least partially explained by the relatively lack of knowledge that we still have regarding the molecular mechanisms underlying them, as well as by the complexity of their etiopathology. In fact, a complex number of changes in the physiological function of the cell has been described in all these aging-associated pathologies, including neurodegenerative disorders. Based on multiple scientific manuscripts produced by us and others, it seems clear that mitochondria are dysfunctional in many of these aging-associated pathologies. For example, mitochondrial dysfunction is an early event in the etiopathology of all the main neurodegenerative disorders, and it could be a trigger of many of the other deleterious changes which are present at the cellular level in these pathologies. While mitochondria are complex organelles and their regulation is still not yet entirely understood, inorganic polyphosphate (polyP) could play a crucial role in the regulation of some mitochondrial processes, which are dysfunctional in neurodegeneration. PolyP is a well-preserved biopolymer; it has been identified in every organism that has been studied. It is constituted by a series of orthophosphates connected by highly energetic phosphoanhydride bonds, comparable to those found in ATP. The literature suggests that the role of polyP in maintaining mitochondrial physiology might be related, at least partially, to its effects as a key regulator of cellular bioenergetics. However, further research needs to be conducted to fully elucidate the molecular mechanisms underlying the effects of polyP in the regulation of mitochondrial physiology in aging-associated pathologies, including neurodegenerative disorders. With a significant lack of therapeutic options for the prevention and/or treatment of neurodegeneration, the search for new pharmacological tools against these conditions has been continuous in past decades, even though very few

P. Urquiza · M. E. Solesio (✉)
Department of Biology, Rutgers University, Camden, NJ, USA
e-mail: m.solesio@rutgers.edu

W. E. G. Müller et al. (eds.), *Inorganic Polyphosphates*, Progress in Molecular and Subcellular Biology 61, https://doi.org/10.1007/978-3-031-01237-2_3

therapeutic approaches have shown potential in treating these pathologies. Therefore, increasing our knowledge about the molecular mechanisms underlying the effects of polyP in mitochondrial physiology as well as its metabolism could place this polymer as a promising and innovative pharmacological target not only in neurodegeneration, but also in a wide range of aging-associated pathologies and conditions where mitochondrial dysfunction has been described as a crucial component of its etiopathology, such as diabetes, musculoskeletal disorders, and cardiovascular disorders.

Keywords Inorganic polyphosphate · PolyP · Neurodegeneration · Bioenergetics · Oxidative phosphorylation · Mitochondrial dysfunction · Mitochondrial unfolded protein response · Mitochondrial permeability transition pore · Stress response · Aging

3.1 Introduction

Our population is aging, this trend is expected to continue in the coming decades (UN 2019). Consequently, the number of individuals suffering from aging-associated pathologies is already exponentially rising. While cellular senescence is a crucial component in the etiopathology of all these aging-associated pathologies, the exact molecular mechanisms underlying senescence and the onset of these pathologies are still far from being elucidated. This is specially the case for the two main neurodegenerative disorders; Alzheimer's Disease (AD) and Parkinson's Disease (PD). These pathologies are currently the first and the second causes of neurodegeneration, respectively. They share some common features at the cellular and histological levels, such as an increased presence of aggregated amyloids in varied tissues, including the brain of patients (Murphy and LeVine 2010; Spillantini et al. 1997). While AD mostly affects the cognitive abilities, PD is mainly a movement disorder, affecting motor functions. Additionally, PD usually presents with dementia in the advanced stages of the disease (Aarsland et al. 1996, 2003; Hely et al. 2008). Specific genetic mutations associated to the familial forms of both diseases have already been described, but most cases of AD and PD are considered sporadic, with an unknown and multifactorial etiology (Bertram and Tanzi 2004; Sandor et al. 2017). In these cases, it seems that aging, specifically cellular senescence, is a crucial contributor to the onset of these disorders. However, we are still far from fully understanding the etiopathology of these disorders, this lack of knowledge makes finding valid treatments extremely challenging. In the United States, neurodegenerative disorders have shown to disproportionally affect underrepresented minorities, including African Americans and Hispanics, when compared to non-Hispanic whites, this is also true for patients belonging to unserved communities (Association 2020, 2021). While the cause(s) underlying these observational studies still remains the object of debate, unequal access to the health system is a likely explanation. Moreover, the presence of aging-associated pathologies, including neurodegenerative disorders, causes not only an emotional, but also a financial

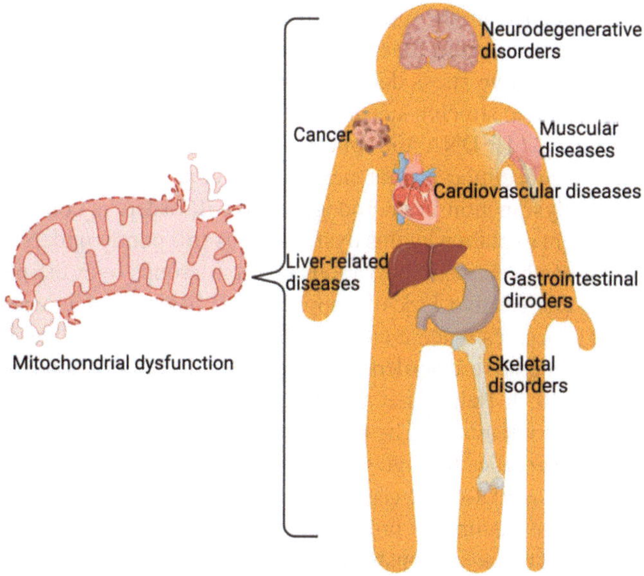

Fig. 3.1 Mitochondrial dysfunction is present in the etiopathology of many aging-associated pathologies. Dysregulated bioenergetics and increased oxidative stress, among other deleterious features, are present in a wide range of aging-associated pathologies. These pathologies affect various organs and systems in the human body. While the exact mechanisms underlying mitochondrial dysfunction in aging remain still mostly unknown, the search for common mechanisms explaining the dysfunction of the organelle in all these pathologies could pave the way for the usage of mitochondrial metabolism as a valid pharmacological target in aging and aging-associated pathologies

burden for the families and loved ones of the affected. In fact in 2020, it was estimated that the total lifetime cost for someone with dementia is $373,527 (Association 2020, 2021). Therefore, understanding the mechanisms underlying these pathologies and finding pharmacological strategies against aging-associated pathologies, including AD and PD, will not only improve the health spam of millions of citizens, but it could help in the fight to reduce inequality (Fig. 3.1).

While the etiopathology of sporadic AD and PD remains mostly unknown, mitochondrial dysfunction has been broadly described as an early feature and a trigger in the deleterious cascade of events which are present at the cellular level in these pathologies (Perier and Vila 2012; Wang et al. 2014). Mitochondria are complex organelles, whose main physiological function in mammalian cells is energy production, they do this by generating ATP in the electron transfer chain (ETC), which is the site of the oxidative phosphorylation (OXPHOS). OXPHOS is the major source of ATP in mammalian cells, even if other extramitochondrial pathways, such as glycolysis, also have the ability to produce this crucial metabolite. An unavoidable byproduct of OXPHOS is the generation of reactive oxygen species (ROS) (Zorov et al. 2014). ROS are crucial components of mammalian cell signalling. The overproduction of these reactive molecules and decreased activity and/or presence of antioxidants, are the source of increased oxidative stress, which is found in all the neurodegenerative disorders, as well as in cellular senescence (Liguori et al. 2018). Increased oxidative stress deleteriously affects a wide variety of mitochondrial and other cellular structures, including (but not limited to) the stability of mitochondrial and cellular DNA, the structure of proteins, and the structural stability of the cellular and subcellular membranes (Liguori et al. 2018).

Mitochondrial dysfunction in neurodegenerative disorders is complex and multifactorial, this might explain why this dysfunction is still poorly understood, despite the progress that has been made by various research groups over the past decades. Notwithstanding this relative lack of knowledge, it has been demonstrated that a good number of processes that are closely related to mitochondria are dysregulated in many neurodegenerative disorders, including AD and PD (Abou-Sleiman et al. 2006; Wang et al. 2014; Baltanás et al. 2013). This includes, for example, bioenergetics and calcium homeostasis. The homeostasis of calcium is intimately related to mitochondrial bioenergetics through the activation of the mitochondrial dehydrogenases. This leads to increased levels of NADH and ATP generation (McCormack et al. 1990). Another example is protein homeostasis, including amyloid management. In fact, it has been demonstrated that misfolded cytoplasmic proteins are imported into mitochondria, most likely for degradation (Ruan et al. 2017). However, the mechanisms involved in this degradation once the proteins are imported into mitochondria still remain very poorly understood. Moreover, the ultimate cause of neuronal cell death in neurodegenerative disorders is the increased presence of apoptosis in neuronal populations, as well as in other cellular populations, in the human brain (Mattson 2000). Apoptosis is a well-known type of programmed cell death. Interestingly, mitochondria play a key role in the activation of the intrinsic pathway of apoptosis, which is the most common pathway of this type of cell death (Wang and Youle 2009). For example, the opening of the mitochondrial permeability transition pore (mPTP) is a point of no return in apoptosis (Gogvadze et al. 2009).

While the deleterious relationship between mitochondrial dysfunction and the onset of neurodegenerative disorders is now clear and well-documented, the exact molecular mechanisms driving and coordinating all these processes that we have mentioned in the previous paragraphs are still not well-understood. As it will be documented in this chapter, it is known that inorganic polyphosphate (polyP) plays an important role in maintaining mitochondrial physiology, and therefore, the regulation of bioenergetics, protein homeostasis, and apoptosis within the organelle (Cremers et al. 2016; Gray et al. 2014; Lempart and Jakob 2019; Pavlov et al. 2010; Solesio et al. 2016a, b, 2020, 2021; Guitart-Mampel et al. 2022; Hambardikar et al. 2022). The intriguing possibility that various authors have proposed and demonstrated in diverse organisms, is that this polyP could be a crucial contributor to the stress response in mammals. Subsequently, this polymer could regulate mitochondrial physiology and contribute toward preventing the dysfunction of the organelle in neurodegenerative disorders, including AD and PD.

3.2 Inorganic Polyphosphate

PolyP is a multifunctional polymer that is highly negatively charged. It is constituted of several orthophosphates, held together by high-energy phosphoanhydride bonds, comparable to those found in ATP (Müller et al. 2017b) (Fig. 3.2). It is impossible to introduce polyP without mentioning the scientific work of the scientist whom most consider the father of the modern study of this polymer. This is the late Nobel laureate Dr. Arthur Kornberg, he was the first researcher to prove that polyP is not merely a molecular fossil, but a crucial component of the stress response in various organisms that ranged from bacteria to mammals (Kornberg 1995; Kornberg et al. 1999; Rao et al. 2009; Rao and Kornberg 1996, 1999). His team, for example, reported that bacteria that show decreased levels of the polymer have an increased sensitivity to different stressors, including heat shock and specific chemicals (Kornberg et al. 1999; Rao et al. 2009).

The study of polyP, which is present in all studied organisms (Denoncourt and Downey 2021), dates back to the late nineteenth century. However, the exact role and the metabolism ancient polymer is still poorly understood in mammals, even if it has been defined as the second most (just after DNA) common negatively charged polymer in living entities. Among the reasons for the lack of knowledge about polyP, we can enumerate the relatively low concentration of polyP in mammalian cells; and the lack of any singular molecular features which could be used for the development of specific biochemical, imaging, or other types of tools allowing for its unequivocal identification. In fact, the different structural conformations in which the polymer can be present in mammals, its widespread cellular distribution, and the ability of polyP to bind different cations and proteins are limiting factors to characterize or analyze polyP in mammals. For example, polyP is a negatively charged molecule, but other biological polymers also share this feature. Moreover, polyP can be present in different lengths (from a few phosphates to long chains of several hundreds). The length of the chain and its structure affects the behavior of the polymer, as it also affects the electronegativity of the molecule, which makes it more difficult to identify. Furthermore, it has been described that polyP can be present free or in association with proteins or other elements, such as membrane proteins or calcium ions (Christ et al. 2020; Solesio et al. 2016a, 2020). Another explanation for the complexity of its identification and studies is the fact that polyP is ubiquitous, even if

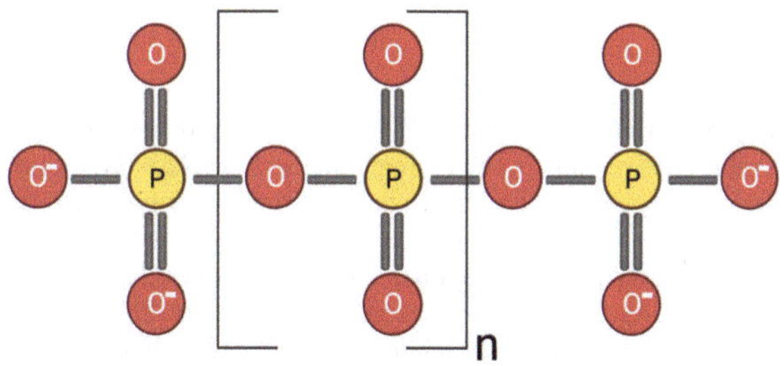

Fig. 3.2 Molecular structure of polyP. PolyP is a polymer that is composed of several (n) orthophosphates, linked together by highly energetic phosphoanhydride bonds. Due to its molecular structure, polyP is a highly negatively charged molecule

we and others have proven a preference for mitochondria, in mammalian cells (Abramov et al. 2007; Angelova et al. 2014; Baev et al. 2017, 2020; Solesio et al. 2016a). In the obituary of Dr. Kornberg, which was published by the New York Times on October 28th, 2007, we found another explanation for why research in the field of polyP is still very challenging; "He (Dr. Kornberg) complained bitterly, however, that too few scientists studied polyphosphate, largely, he said, because of science's proclivity to work "in a clannish way." With more scientists struggling for grants in an era of tight budgets, he said, "nobody is going to propose doing anything that is bold or creative," like working on polyphosphate.

Despite all these difficulties, some tools are already available for the polyP research community (Solesio and Pavlov 2016). For example, while the metabolism of the polymer also remains mostly unknown in mammalian cells, thanks to the scientific work of Dr. Kornberg and other colleagues, this metabolism is already quite well-described in prokaryotes and lower eukaryotes, including bacteria and yeast (Ahn and Kornberg 1990; Akiyama et al. 1992, 1993; Ishige et al. 2002; Kumble and Kornberg 1996). Thanks to these studies, some authors were able to develop the very few molecular tools which are currently available for the study of the role of polyP in mammalian systems and disorders, including in neurodegeneration. These tools include the expression, in mammalian cells, of highly conserved enzymes from bacteria and yeast, which are involved in the metabolism of polyP, such as the endopolyphosphate (PPN) or exopolyphosphatase (PPX) (both are enzymes that mediate the non-processive and the processive cleavage of polyP, respectively); or the polyP kinase (PPK), which is involved in the biosynthesis of polyP through the catabolism of ATP (Bondy-Chorney et al. 2020; Solesio et al. 2021; Guitart-Mampel et al. 2022; Hambardikar et al. 2022) (Fig. 3.3). Additionally, a few other strategies are also available to detect or quantify polyP. These include [^{32}P] radiolabeling (Cowling and Birnboim 1994; Szewczyk et al.

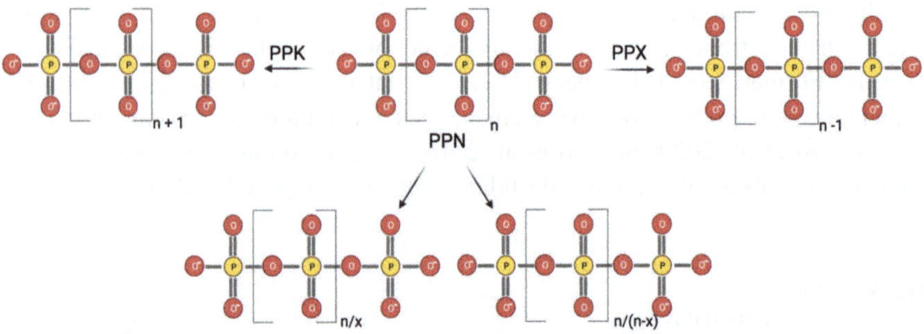

Fig. 3.3 Metabolism of polyP. While the metabolism of polyP remains still mostly unknown in mammalian organisms, including humans, it is relatively well-described in other simple organisms, including bacteria and yeast. In these organisms, polyP is elongated by adding phosphate groups to the free end of the polymer. This reaction is catalyzed by the polyphosphate kinase (PPK) enzyme. Moreover, the polymer is reduced by cleaving phosphates from its free ends. The enzyme in charge of catalyzing this process is exopolyphosphatase (PPX). Lastly, the chains of polyP can also be cleavaged internally by the endopolyphosphatase enzyme (PPN)

1978); the use of specific dyes which interact with polyP, such as Toluidine blue O or fluorescent dye 4′,6-diamidino-2-phenylindole (DAPI) (Aschar-Sobbi et al. 2008; Mullan et al. 2002; Ohtomo et al. 2008; Streichan et al. 1990); and nuclear magnetic resonance (NMR) spectroscopy (Bental et al. 1991; Christ et al. 2019). However, these methods are far from perfect. For example, [^{32}P] radiolabeling only allows for the detection of lower amounts of polyP (Desfougeres et al. 2020). Additionally, free phosphates or even other negative molecules, such as DNA or RNA, are difficult to distinguish from polyP with some of the current analytical techniques (Christ et al. 2020). Lastly, some of these techniques require the use of expensive equipment like NMR, and some others, like radiolabeling need to be conducted in specific facilities for safety reasons. Despite all these problems, the polyP research community, which is small but highly collaborative, has been working tirelessly to elucidate the functions of polyP.

Of special interest for this chapter are the results obtained from some research groups that focused their studies on the study of the effects of polyP in the stress response in mammalian cells, and more specifically in mammalian mitochondria. This stress response is a common response to diverse stressors, including the onset of neurodegenerative disorders. The current scientific consensus in the field states that in mammalian cells polyP is composed of a few orthophosphate units, and the concentrations of the polymer are within the micromolar range (Kumble and Kornberg 1995). An example of this is present in the brains of rodents, where polyP concentrations range up to 100 μM (Gabel and Thomas 1971; Kumble and Kornberg 1995). Moreover, we and others have shown the preference for polyP to localize within mitochondria in these organisms (Baev et al. 2017, 2020; Borden et al. 2021; McIntyre and Solesio 2021; Müller et al. 2019; Pavlov et al. 2010; Solesio et al. 2016a, b, 2020). Interestingly, supporting the studies conducted by Dr. Kornberg, other researchers have demonstrated that differences in the intracellular concentration of polyP associate with age and disease, emphasizing the likely effect of this polymer in the stress-associated cellular response, comparable to the stress response found in neurodegenerative disorders (Angelova et al. 2014). It is of note that increased cellular stress, especially when this is induced by increased oxidative stress, is a trigger of dysregulated bioenergetics in diverse mammalian cells and tissues (Armstrong et al. 2018; Judge and Leeuwenburgh 2007; Paradies et al. 2010). However, the exact mechanisms that link these two processes are still not yet totally understood.

3.3 PolyP and Bioenergetics

The exact meaning of the term bioenergetics is complex and contentious. Nowadays, the current definition of the term is expanded to include OXPHOS and glycolysis, to encompass all the mechanisms by which the cellular production and demand for ATP are matched and scaled to meet the needs of the mammalian cell, under various states of demand. OXPHOS and glycolysis and their regulation are well-known and interconnected processes in this characterization of bioenergetics (Spinelli and

Haigis 2018). However, other pathways and key molecules may also be involved in the regulation of this network to provide the required flux of ATP. The mechanism of the regulation of these processes remains mostly unknown, particularly under stress conditions, including in neurodegeneration.

Mitochondria are crucial organelles for the maintenance of mammalian physiology, including human bioenergetics. While other extramitochondrial pathways are also involved in this process, the majority of ATP is produced within the organelle in most of the tissues of these organisms, under physiological conditions. Moreover, mitochondrial dysregulation, including bioenergetics dysfunction, is an early step and trigger in the etiopathology of many neurodegenerative disorders, including AD and PD (Requejo-Aguilar and Bolanos 2016; Sonntag et al. 2017). Interestingly, polyP can be found in various subcellular locations in diverse eukaryotic organisms, including the membrane (associated with proteins), nucleus, cytoplasm, and other organelles (Kumble and Kornberg 1995), as well as in the extracellular space (Suess et al. 2017). This wide range of locations seems to suggest (depending on its location), different roles for polyP in cellular metabolism. In fact, in mammalian cells, polyP is highly co-localized to mitochondria (Abramov et al. 2007; Solesio et al. 2016a, 2020), which seems to indicate a role for the polymer in maintaining mitochondrial bioenergetics, as this is the main function of the organelle in the human metabolism. Accordingly, the crucial role played by polyP in the regulation of bioenergetics has been broadly demonstrated by a number of different research groups (Abramov et al. 2007; Freimoser et al. 2006; Müller et al. 2017a, b; Seidlmayer et al. 2015, 2019; Solesio et al., 2016a, 2020; Suess et al. 2017). For example, in different mammalian cells, including kidney and adrenal cells, and fibroblasts; the close relationship between cellular levels of ATP and polyP has been proven (Kumble and Kornberg 1995). Furthermore, our previous work demonstrated the important regulatory effects of mitochondrial polyP in the maintenance of the levels of free calcium within the organelle in mammalian cells (Solesio et al. 2016b, 2020). As previously mentioned, intramitochondrial calcium is crucial in the production of ATP by activating the dehydrogenase enzymes within the organelle (McCormack et al. 1990). Moreover, our recent publications shows the crucial role played by mitochondrial polyP in preventing the shift from OXPHOS to glycolysis in human cell lines (Solesio et al. 2021; Guitart-Mampel et al. 2022), as well as on the regulation of the production of ROS (Hambardikar et al. 2022). Additionally, in an elegant study conducted in mammalian cellular models, the authors showed that exogenous polyP is able to reverse the compromised energy status of the cells that were induced by treatment and the aggregation of the amyloid peptide (Müller et al. 2017a), which is a well-known disturber of cellular bioenergetics. The increased presence of aggregated amyloids in the brain is a common feature of neurodegenerative disorders, including AD and PD (Murphy and LeVine 2010; Spillantini et al. 1997). In fact, corroborating this study, another group of authors used multiple mammalian cell lines, including differentiated SH-SY5Y, PC12, and HeLa, to show the protective effect of polyP against the toxicity exerted by the amyloid-β peptide (Canevari et al. 2004). The authors hypothesize that the protective effect could be directly exerted by the regulatory effects of polyP on the regulation of bioenergetics, or by its action as a molecular chaperone.

Furthermore, in mammals it has been established that polyP levels are not fixed, but volatile and highly associated with the metabolic state of mitochondria (Pavlov et al. 2010). In fact, mammalian cells that were treated with substrates of mitochondrial respiration showed increased levels of polyP, the opposite effect was demonstrated when cells were treated with inhibitors or uncouplers of mitochondrial respiration. Moreover, oligomycin, which is a well-known inhibitor of the mitochondrial F_0F_1-ATP synthase (Salomon et al. 2000), completely blocked the production of polyP. Some evidence in support of these findings is that another study has recently proved that the mitochondrial F_0F_1-ATP synthase plays a crucial role in the metabolism of polyP (Bayev et al. 2020), which for the most part is still unknown, especially in mammalian cells. The authors demonstrated that polyP could be both hydrolyzed and produced by the F_0F_1-ATP synthase, which can be stimulated by the polymer in the presence of ATP. Lastly, in the context of PD, increased intracellular levels of polyP were found in LRRK2 and Pink1 (leucine-rich repeat kinase 2 and PTEN induced kinase 1, respectively) knockout cells (Angelova et al. 2014). Mutations in the genes that code for these proteins result in some of the most common familial forms of PD (Nuytemans et al. 2010). Mitochondrial dysfunction has also been broadly described in the familial forms of PD. In fact, the main genes which are mutated in this pathology are directly linked to mitochondrial functions. These genes include Pink1, Parkin, or LRRK2 (Kilarski et al. 2012). While other studies show a protective effect of polyP in preventing mitochondrial dysfunction, the results of this manuscript seem to point in the opposite direction. One explanation for these findings could be that the observed increments in the levels of polyP could, however, not be sufficient to counteract the immense bioenergetics dysregulation which is present in the familial cases of PD. These data indicate the crucial direct or indirect regulatory role of polyP in mammalian bioenergetics. However, the exact molecular mechanism(s) involved in this regulation still remains largely a mystery.

3.4 Mitochondrial Physiology and PolyP in Neurodegeneration

A typical consequence of dysregulated bioenergetics is mitochondrial dysfunction, which triggers a deleterious vicious cycle that, ultimately, induces increased apoptotic cell death (Fig. 3.4). This sequence of events has been broadly demonstrated in the main neurodegenerative disorders, AD and PD. In fact, the higher degree of the dysregulation of bioenergetics in the organelle, the more affected the main components of mitochondrial physiology become, which increases, even more, the dysregulation of bioenergetics. Indeed, the interrelationship between dysregulated mitochondrial bioenergetics and the dysregulation of the two main quality control

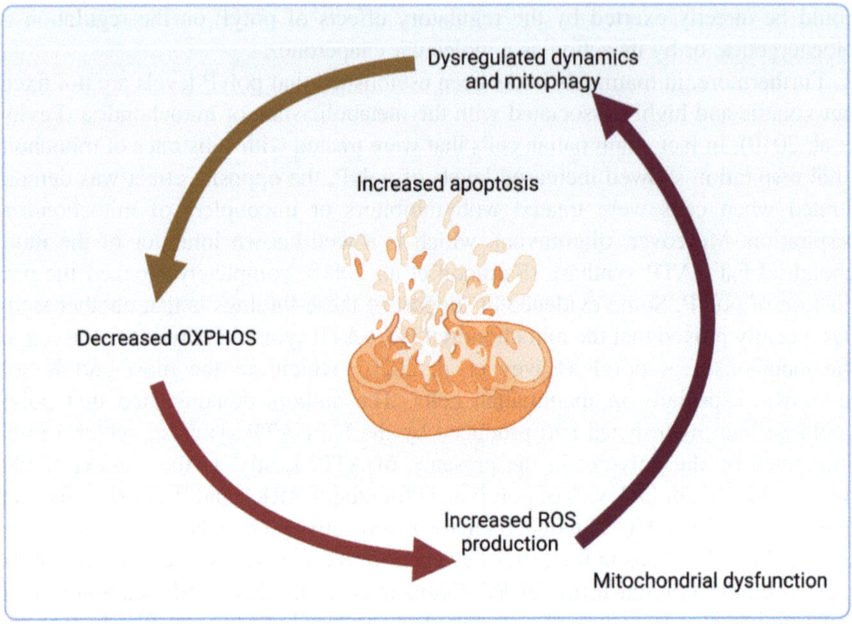

Fig. 3.4 Dysregulated mitochondrial function is the ultimate trigger of increased apoptotic cell death. In many aging-associated pathologies, including neurodegenerative disorders, increased apoptosis has been broadly demonstrated in various tissues and organs. While the triggers of this increase are complex and multifactorial, the ultimate activator of apoptosis is the increased dysfunction of mitochondria. This dysfunction has different components, which are interconnected in a deleterious cycle that ultimately increases, even more, the degree of the dysfunction. Among these components, dysregulated mitochondrial dynamics and mitophagy, dysfunctional bioenergetics, and increased generation of ROS have been demonstrated as crucial contributors toward increased apoptosis in neurodegenerative disorders

mechanisms in the organelle, mitophagy, and mitochondrial Unfolded Protein Response (UPR^{mt}), has been broadly described in a wide variety of organisms, including mammals (Melber and Haynes 2018; Poirier et al. 2019; Priault et al. 2005; Schulz and Haynes 2015), even if the mechanisms underlying this relationship still remain mostly unknown. Mitophagy and UPR^{mt} have also been shown to play a critical role in the maintenance of protein homeostasis, a process in which chaperones are crucial components.

Dysregulated mitochondrial bioenergetics also has extramitochondrial consequences in the mammalian cell, including the shift from OXPHOS to glycolysis (Warburg effect; Chen et al. 2015). This effect is a typical characteristic of cancer cells, but it has also been proven in many other mammalian cells and systems (Chen et al. 2015). Interestingly, the regulatory role played by the levels of mitochondrial polyP in this process is known, it was elucidated by our previous work (Solesio et al. 2021). In that research article, we also demonstrated that mammalian mitochondria that lack polyP in the organelle are smaller than the control mitochondria. While decreased size is often a hallmark of increased rates of mitochondrial fission, our data did not show increased phosphorylation of Drp1, which is needed for mitochondrial fission (Chang and Blackstone 2007). Increased mitochondrial fission is also a classical consequence and trigger of dysregulated bioenergetics in mammalian cells (Westermann 2012). One plausible explanation for the lack of increased phosphorylation of Drp1 in our models could be found in the effects of polyP on the phosphorylation status of the cells. Using model organisms, it has been demonstrated that polyP is involved in the polyphosphorylation of varied proteins (Azevedo et al. 2015; Azevedo et al. 2018). This could likely affect the effects of polyP in the regulation of mitochondrial physiology, both under control conditions and in the presence of increased stress, like in main neurodegenerative disorders.

Despite the efforts made in the past decades toward elucidating the exact molecular structure of the mPTP, this still remains elusive. Interestingly, some researchers have demonstrated the crucial role exerted by polyP as a structural component of the mPTP, as well as in the regulation of the formation and the opening of this pore, which is key for mitochondrial physiology and cell fate (Seidlmayeret al. 2012a, b, 2019; Wang et al. 2003). The effects of polyP in the regulation of the mPTP provide us with further bibliographical support, showing the important role played by polyP in the regulation of mitochondrial physiology. The studies cited above have been conducted in various organism models and cell types, including primary mammalian cardiomyocytes and mitoplasts. Interestingly, the close relationship between the mPTP and the levels of mitochondrial calcium is also well-documented. This relationship is explained by the effects that the formation and the opening of the mPTP have on the mitochondrial membrane potential (Webster 2012). In fact, increased intramitochondrial calcium is a well-known trigger of the formation and the opening of the mPTP (McCormack et al. 1990; Nicholls 1978). Consequently, the formation and the opening of this structure are also affected by the bioenergetics status of cells, which as previously mentioned, is dysfunctional in neurodegeneration and regulated, at least partially, by polyP (Elustondo et al. 2016). The role exerted by polyP in the regulation of the mPTP will certainly also affect the rates of neuronal apoptotic cell death. These rates are clearly increased in neurodegenerative disorders, including AD and PD (Mattson 2000). Further support for the regulatory role of polyP on apoptosis has been provided by research in which the authors demonstrated the role of polyP in cancer progression, a disease that is characterized by inhibition of apoptosis (Wong 2011). Specifically, using human MCF-7 breast cancer or myeloma, it has been shown that increased amounts of polyP reduced cell growth and/or induced apoptosis (Hernandez-Ruiz et al. 2006; Wang et al. 2003).

PolyP has also been shown to be involved in other mitochondrial processes which are crucial for the physiology of the organelle, even if the exact molecular mechanisms of this involvement remain, once again, mostly unknown. These processes include the stimulation of the mammalian target of rapamycin (mTOR), which is a protein kinase that is involved in mitochondrial physiology. Dysregulated metabolism of mTOR has been demonstrated in varied pathologies, including AD and PD (Santini et al. 2009). Significantly, polyP has also been shown to have potent chaperoning effects. Interestingly, this ability has been proven to also be efficient when amyloids; specifically, with αSyn, which is the main amyloid present in PD; similar to those found in neurodegeneration, are present (Cremers et al. 2016; Gray et al. 2014; Lempart and Jakob, 2019; Lempart et al. 2019; Yoo et al. 2018). The chaperoning effects of polyP in the amyloidogenic process are exerted by its ability to stabilize scaffold β-sheet-containing protein-folding intermediates (Lempart and Jakob 2019), and it has also been demonstrated in nonmammalian organisms, including *Escherichia coli* and *Caenorhabditis elegans* (Cremers et al. 2016; McColl et al. 2012). Other researchers have also proven the regulatory role of polyP in non-amyloidogenic proteins, such as collagen (Khong et al. 2020), as well as the protective effects of polyP in protein homeostasis in bacteria against a plethora of stress inductors, including oxidative, thermal, and chemical stresses (Alcantara et al. 2014; Jahid et al. 2006; Nikel et al. 2013; Yoo et al. 2018). However, here too, the exact molecular mechanisms explaining the effects of polyP in the regulation of all these processes, and whether these regulatory effects are exerted directly at different levels of mitochondrial physiology or as a consequence of the potent action of the polymer as a component of mammalian bioenergetics remain mostly unknown, especially in mammalian cells and under stress conditions (Fig. 3.5).

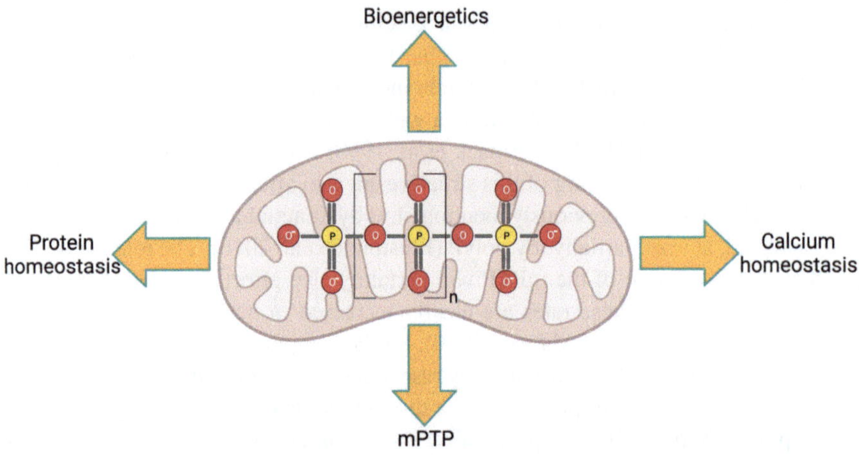

Fig. 3.5 PolyP is involved in the regulation of crucial components of mitochondrial physiology. The key role played by polyP in the regulation of bioenergetics, calcium, and protein homeostasis, as well as in the formation and regulation of the opening of the mPTP has been broadly demonstrated by various authors working on a wide variety of models. All these processes are closely interconnected and crucial for the proper physiological function of mitochondria. Moreover, they are all dysfunctional in neurodegenerative disorders, including AD and PD

3.5 PolyP, Mitochondria, and Other Aging-Associated Disorders

PolyP has also shown a potent effect on a wide variety of physiological functions and processes which are affected in aging and aging-associated pathologies. These include inflammation, blood coagulation, cancer progression, and bone and cartilage tissue development. Although there are a plethora of publications reporting these functions, the lack of knowledge regarding the exact molecular mechanisms underlying these processes, as well as of the metabolism of polyP in mammals, limits the use of this molecule as a valid pharmacological target in aging-associated pathologies.

For example, several studies revealed that polyP is a potential modulator in the plasma clotting cascade, and consequently in hemostasis, inflammation, and thrombosis (Baker et al. 2019; Morrissey et al. 2012; Smith and Morrissey 2014; Travers et al. 2015; van der Meijden and Heemskerk 2019). Specifically, it has been demonstrated that the activation of human platelets secretes granules that are enriched with different small molecules of a diverse nature. These molecules include but are not limited to polyP, ADP, serotonin, glutamate, histamine, and calcium; and they are crucial for maintaining proper hemostasis (Koupenova et al. 2018). The secreted polyP from the activated platelets has a length between 60 and 100 phosphates (Baker et al. 2019), and contributes toward the activation of factor V, which is one of the main clotting factors and an upstream member of the coagulation cascade (Duga et al. 2004). Interestingly, affected hemostasis, including coagulation ability of the organism, has been described in aging and aging-associated pathologies (Mari et al. 2008). Moreover, the crosstalk between coagulation and inflammation is well-known (Koupenova et al. 2018; Maas and Renne 2018; Oikonomou et al. 2020). While the effects of polyP on coagulation have been intensively investigated, the mechanisms of its proinflammatory action remain still poorly understood. This action has been demonstrated in some studies, which showed that the platelets released-polyP amplifies the proinflammatory response and activates the NF-κB signalling pathway, which ultimately induces the expression of proinflammatory genes (Baker et al. 2019; Hassanian et al. 2017; Liu et al. 2017).

Although platelets only have 5–8 mitochondria per cell, they are highly metabolically active cells (Holmsen 1985). In these cells, it has been demonstrated that mitochondria supply 50% of the total required energy (Aibibula et al. 2018; Baccarelli and Byun 2015; Doery et al. 1970). Moreover, it is also known that mitochondrial dysfunction in platelets, similarly to what happens in other mammalian cells, increases the levels of oxidative stress, which leads to cell damage and apoptosis (Sinauridze et al. 2007; Swerdlow et al. 1997). This pathological scenario is especially critical in neurodegeneration (Kocer et al. 2013), as platelets are the second highest (just after neurons) reservoir of amyloid precursor protein (APP) in the human body. APP in platelets contributes to increasing the aggregated levels of amyloids that circulate in blood (Bush et al. 1990). Interestingly, a study which was conducted on patients suffering from chronic thrombocytopenia, showed increased levels of polyP in the platelets of patients suffering from this condition, when compared to healthy individuals (Zurawska-Plaksej et al. 2020). This increase could be explained as a compensatory mechanism to prevent excessive bleeding, and it could also affect APP in circulating blood.

As previously mentioned, polyP is also involved in the formation and/or regeneration of cartilage and bone tissue (Hacchou et al. 2007; Omelon et al. 2009; Schröder et al. 2000). This process is likewise affected in aging-associated pathologies (Boskey and Coleman 2010; Lotz and Loeser 2012). The studies of the effects of polyP in bone and cartilage were conducted using exogenous polyP and human cultured cells which are relevant for systems such as mesenchymal stem cells (MSCs), osteoblasts cells, and osteoclasts cells (Wang et al. 2019). MSCs cells are multipotent cells that are localized in almost every type of connective tissue. However, they show an especially high presence in the bone marrow. MSCs can be differentiated into specialized cells, including osteoblasts, adipocytes, and chondrocytes (Uccelli et al. 2008). The differentiation of MSCs is closely controlled by mitochondrial physiology (Li et al. 2017) and the status ofmitochondria is crucial in the physiology of chondrocytes (Castro et al. 2020). Accordingly, undifferentiated MSCs show low mitochondrial activity, while after activation, they show increased mitochondrial biogenesis and OXPHOS (Hofmann et al. 2012; Hsu et al. 2016; Tahara et al. 2009). The studies regarding the role of polyP which were conducted in MSCs demonstrate that polyP promotes the differentiation and maturation of MSCs into osteoblast (Bruedigam et al. 2010), which concurs with the crucial role played by polyP on mitochondrial physiology. Moreover, in the presence of calcium, polyP accelerates the proliferation and mineralization of osteoblasts (Kawazoe et al. 2004, 2008; Müller et al. 2011). Osteoblasts and osteoclasts are specialized cells that are involved in the mineralization of bones and in the resorption and modelling or remodeling of bones, respectively (Dirckx et al. 2019).

While polyP induces the activity of osteoblasts, the exact function of the interaction between polyP and osteoclasts still remains mostly unknown, as the studies of its effects are diffuse and strongly depend on the length of the polymer. However, it seems clear that the degradation of polyP by osteoclasts improves the environment during bone resorption (Harada et al. 2013). Additionally, the development of modelling and remodeling of bones demands high levels of energy. Accordingly, it is known that mature osteoblasts accumulate mitochondria, which increases energy generation. The increased ATP mostly is used for protein synthesis, including the synthesis of osteocalcin, which is required for several physiological processes (Dudley and Spiro 1961). Osteocalcin, which is a protein synthesized by osteoblasts and considered the main marker of bone formation (Wei et al. 2015), is also involved in the physiology of dopaminergic neurons, which are the main type of neurons affected in PD (Maetzler and Berg 2018). Moreover, decreased levels of dopamine within the brain have also been shown to increase the risk of suffering AD (Pan et al. 2019). Interestingly, it has been documented that osteocalcin crosses the blood–brain barrier (BBB), and binds to neurons in the brain to facilitate the synthesis of dopamine (Khrimian et al. 2017; Oberle and Stahl 1990). In fact, some studies correlate the status of bone mineralization density with the incidence of neurodegenerative disorders, including AD and PD (Bradburn et al. 2016; Shan et al. 2019). Furthermore, studies conducted using PD animal models which were treated with osteocalcin, showed that this protein protected against dopaminergic neuronal damage, resulting in improved motor function (Guo et al. 2018). Lastly, the interaction of polyP is also demonstrated in cartilage formation. The effects of polyP are chain length- and concentration-dependent, in both cell cultures and ex vivo cultures of cartilage tissue (St-Pierre et al. 2012). This ability of polyP has been used by biomaterials engineers for the design of scaffolds for cartilage tissue. Because cartilage are connective tissue that do not regenerated when damaged and are affected in aging, the biophysical properties of polyP-based materials are expected to have an important impact in this field (Wang et al. 2019).

Lastly, polyP is also a gliotransmitter in astrocytes (Holmstrom et al. 2013), which can release polyP after calcium stimulation; it is then taken up by neurons (Holmstrom et al. 2013). The crucial role played by the dysfunction of glial cells, including astrocytes, in AD and PD has been broadly described (Dzamba et al. 2016; Vila et al. 2001). Although the exact role of polyP in neuronal signalling remains unclear, the available data suggests that polyP as a gliotransmitter may have an important implication in cognition and motor behaviors.

3.6 Outlook and Conclusion

PolyP is present in many different mammalian compartments, including fluids such as synovial fluid, blood; and specialized cells (e.g., platelets, fibroblasts, and osteoblasts). Its cellular function seems to be closely related to the concentration of the polymer in each subcellular location, as well as to the chain length. The high co-localization between mitochondria and polyP in mammals hints toward a role for the polymer in the regulation of bioenergetics. This role, as well as the effects of polyP on the maintenance of other crucial aspects of mitochondrial physiology, has been broadly demonstrated. Interestingly, mitochondrial dysfunction, including the dysregulation of the processes in which the regulation of polyP is involved, is a well-known component in aging-associated pathologies, including the two main neurodegenerative diseases, AD and PD. Specifically, the dysregulation of mitochondrial physiology includes but is not limited to the dysfunction of bioenergetics and protein homeostasis. Mitochondrial physiology is obviously closely interconnected with cell physiology and its dysregulation has disastrous effects on cellular survival, including increased apoptosis, which is one of the main hallmarks of neurodegeneration.

However, the exact molecular mechanism underlying the action of polyP in the regulation of mitochondrial physiology remains still mostly unknown. In fact, one of the principal questions in the field, currently, is whether the effects of polyP in the regulation of bioenergetics are somehow underlying the rest of the effects of the polymer observed in the physiology of the organelle, or can polyP exert a plethora of effects on multiple mitochondrial and cellular levels. The main reason for this lack of knowledge about polyP is the relative absence of specific experimental tools that allow for the assaying of its levels and subcellular location, at different times. Additionally, the fact that the metabolism of the polymer still remains unclear, especially in mammalian cells, including human cells, adds an additional level of complexity to the study of this polymer. It is in our considered and humble opinion that further research should be conducted to clarify the exact effects of polyP in the regulation of mitochondrial physiology, as well as the molecular mechanisms of these effects. These studies will be of special interest in the field of neurodegeneration, taking into account that preventing mitochondrial dysfunction could be a valid pharmacological target in these conditions, including AD and PD.

Acknowledgments We kindly thank Mr. Mitch Maleki, Esq., for editing this book chapter. The writing of this book chapter was partially supported by the National Institutes of Health (4R00AG055701-03 to MES); by Start Up funds provided by Rutgers University to MES; and by the American Heart Association, Career Development Award (854708 to MES). Images were created using BioRender.com.

References

Aarsland D, Tandberg E, Larsen JP, Cummings JL (1996) Frequency of dementia in Parkinson disease. Arch Neurol 53:538–542

Aarsland D, Andersen K, Larsen JP, Lolk A, Kragh-Sorensen P (2003) Prevalence and characteristics of dementia in Parkinson disease: an 8-year prospective study. Arch Neurol 60:387–392

Abou-Sleiman PM, Muqit MM, Wood NW (2006) Expanding insights of mitochondrial dysfunction in Parkinson's disease. Nat Rev Neurosci 7:207–219

Abramov AY, Fraley C, Diao CT, Winkfein R, Colicos MA, Duchen MR, Pavlov E et al (2007) Targeted polyphosphatase expression alters mitochondrial metabolism and inhibits calcium-dependent cell death. Proc Natl Acad Sci U S A 104:18091–18096

Ahn K, Kornberg A (1990) Polyphosphate kinase from *Escherichia coli*. Purification and demonstration of a phosphoenzyme intermediate. J Biol Chem 265:11734–11739

Aibibula M, Naseem KM, Sturmey RG (2018) Glucose metabolism and metabolic flexibility in blood platelets. J Thromb Haemost 16:2300–2314

Akiyama M, Crooke E, Kornberg A (1992) The polyphosphate kinase gene of *Escherichia coli*. Isolation and sequence of the ppk gene and membrane location of the protein. J Biol Chem 267: 22556–22561

Akiyama M, Crooke E, Kornberg A (1993) An exopolyphosphatase of *Escherichia coli*. The enzyme and its *ppx* gene in a polyphosphate operon. J Biol Chem 268:633–639

Alcantara C, Blasco A, Zuniga M, Monedero V (2014) Accumulation of polyphosphate in *Lactobacillus* spp. and its involvement in stress resistance. Appl Environ Microbiol 80:1650–1659

Angelova PR, Agrawalla BK, Elustondo PA, Gordon J, Shiba T, Abramov AY, Pavlov EV et al (2014) *In situ* investigation of mammalian inorganic polyphosphate localization using novel selective fluorescent probes JC-D7 and JC-D8. ACS Chem Biol 9:2101–2110

Armstrong JA, Cash NJ, Ouyang Y, Morton JC, Chvanov M, Latawiec D, Criddle DN et al (2018) Oxidative stress alters mitochondrial bioenergetics and modifies pancreatic cell death independently of cyclophilin D, resulting in an apoptosis-to-necrosis shift. J Biol Chem 293:8032–8047

Aschar-Sobbi R, Abramov AY, Diao C, Kargacin ME, Kargacin GJ, French RJ, Pavlov E (2008) High sensitivity, quantitative measurements of polyphosphate using a new DAPI-based approach. J Fluoresc 18:859–866

Association A s (2020) 2020 Alzheimer's disease facts and figures. Alzheimers Dement. https://doi.org/10.1002/alz.12068

Association A s (2021) 2021 Alzheimer's disease facts and figures. Alzheimers Dement 17:327–406. https://doi.org/10.1002/alz.12328

Azevedo C, Livermore T, Saiardi A (2015) Protein polyphosphorylation of lysine residues by inorganic polyphosphate. Mol Cell 58:71–82

Azevedo C, Singh J, Steck N, Hofer A, Ruiz FA, Singh T, Saiardi A et al (2018) Screening a protein array with synthetic biotinylated inorganic polyphosphate to define the human polyP-ome. ACS Chem Biol 13:1958–1963

Baccarelli AA, Byun HM (2015) Platelet mitochondrial DNA methylation: a potential new marker of cardiovascular disease. Clin Epigenetics 7:44

Baev AY, Negoda A, Abramov AY (2017) Modulation of mitochondrial ion transport by inorganic polyphosphate - essential role in mitochondrial permeability transition pore. J Bioenerg Biomembr 49:49–55

Baev AY, Angelova PR, Abramov AY (2020) Inorganic polyphosphate is produced and hydrolyzed in F0F1-ATP synthase of mammalian mitochondria. Biochem J 477:1515–1524

Baker CJ, Smith SA, Morrissey JH (2019) Polyphosphate in thrombosis, hemostasis, and inflammation. Res Pract Thromb Haemost 3:18–25

Baltanás A, Solesio ME, Zalba G, Galindo MF, Fortuño A, Jordán J (2013) The senescence-accelerated mouse prone-8 (SAM-P8) oxidative stress is associated with upregulation of renal NADPH oxidase system. J Physiol Biochem 69:927–935

Bayev AY, Angelova PR, Abramov AY (2020) Inorganic polyphosphate is produced and hydrolysed in F_0F_1-ATP synthase of mammalian mitochondria. Biochem J. https://doi.org/10.1042/BCJ20200042

Bental M, Pick U, Avron M, Degani H (1991) Polyphosphate metabolism in the alga *Dunaliella salina* studied by ^{31}P-NMR. Biochim Biophys Acta 1092:21–28

Bertram L, Tanzi RE (2004) Alzheimer's disease: one disorder, too many genes? Hum Mol Genet 13(Spec No 1):R135–141. https://doi.org/10.1093/hmg/ddh077

Bondy-Chorney E, Abramchuk I, Nasser R, Holinier C, Denoncourt A, Baijal K, McCarthy L, Khacho M, Lavallée-Adam M, Downey M (2020) A broad response to intracellular long-chain polyphosphate in human cells. Cell Rep 33:108318

Borden EA, Furey M, Gattone NJ, Hambardikar VD, Liang XH, Scoma ER, Solesio ME et al (2021) Is there a link between inorganic polyphosphate (polyP), mitochondria, and neurodegeneration? Pharmacol Res 163:105211

Boskey AL, Coleman R (2010) Aging and bone. J Dent Res 89:1333–1348

Bradburn S, McPhee JS, Bagley L, Sipila S, Stenroth L, Narici MV, Murgatroyd C et al (2016) Association between osteocalcin and cognitive performance in healthy older adults. Age Ageing 45:844–849

Bruedigam C, Eijken M, Koedam M, van de Peppel J, Drabek K, Chiba H, van Leeuwen JP (2010) A new concept underlying stem cell lineage skewing that explains the detrimental effects of thiazolidinediones on bone. Stem Cells 28:916–927

Bush AI, Martins RN, Rumble B, Moir R, Fuller S, Milward E et al (1990) The amyloid precursor protein of Alzheimer's disease is released by human platelets. J Biol Chem 265:15977–15983

Canevari L, Abramov AY, Duchen MR (2004) Toxicity of amyloid beta peptide: tales of calcium, mitochondria, and oxidative stress. Neurochem Res 29:637–650

Castro CM, Corciulo C, Solesio ME, Liang F, Pavlov EV, Cronstein BN (2020) Adenosine A2A receptor (A2AR) stimulation enhances mitochondrial metabolism and mitigates reactive oxygen species-mediated mitochondrial injury. FASEB J 34:5027–5045

Chang CR, Blackstone C (2007) Cyclic AMP-dependent protein kinase phosphorylation of Drp1 regulates its GTPase activity and mitochondrial morphology. J Biol Chem 282:21583–21587

Chen X, Qian Y, Wu S (2015) The Warburg effect: evolving interpretations of an established concept. Free Radic Biol Med 79:253–263

Christ JJ, Willbold S, Blank LM (2019) Polyphosphate chain length determination in the range of two to several hundred P-subunits with a new enzyme assay and ^{31}P NMR. Anal Chem 91: 7654–7661

Christ JJ, Willbold S, Blank LM (2020) Methods for the analysis of polyphosphate in the life sciences. Anal Chem 92:4167–4176

Cowling RT, Birnboim HC (1994) Incorporation of [^{32}P]orthophosphate into inorganic polyphosphates by human granulocytes and other human cell types. J Biol Chem 269:9480–9485

Cremers CM, Knoefler D, Gates S, Martin N, Dahl JU, Lempart J, Jakob U et al (2016) Polyphosphate: a conserved modifier of amyloidogenic processes. Mol Cell 63:768–780

Denoncourt A, Downey M (2021) Model systems for studying polyphosphate biology: a focus on microorganisms. Curr Genet 67:331–346

Desfougeres Y, Saiardi A, Azevedo C (2020) Inorganic polyphosphate in mammals: where's Wally? Biochem Soc Trans 48:95–101

Dirckx N, Moorer MC, Clemens TL, Riddle RC (2019) The role of osteoblasts in energy homeostasis. Nat Rev Endocrinol 15:651–665

Doery JC, Hirsh J, Cooper I (1970) Energy metabolism in human platelets: interrelationship between glycolysis and oxidative metabolism. Blood 36:159–168

Dudley HR, Spiro D (1961) The fine structure of bone cells. J Biophys Biochem Cytol 11:627–649

Duga S, Asselta R, Tenchini ML (2004) Coagulation factor V. Int J Biochem Cell Biol 36:1393–1399

Dzamba D, Harantova L, Butenko O, Anderova M (2016) Glial cells – the key elements of Alzheimer's disease. Curr Alzheimer Res 13:894–911

Elustondo PA, Nichols M, Negoda A, Thirumaran A, Zakharian E, Robertson GS, Pavlov EV (2016) Mitochondrial permeability transition pore induction is linked to formation of the complex of ATPase C-subunit, polyhydroxybutyrate and inorganic polyphosphate. Cell Death Discov 2:16070

Fonseca TB, Sanchez-Guerrero A, Milosevic I, Raimundo N (2019) Mitochondrial fission requires DRP1 but not dynamins. Nature 570:E34–E42

Freimoser FM, Hurlimann HC, Jakob CA, Werner TP, Amrhein N (2006) Systematic screening of polyphosphate (poly P) levels in yeast mutant cells reveals strong interdependence with primary metabolism. Genome Biol 7:R109

Gabel NW, Thomas V (1971) Evidence for the occurrence and distribution of inorganic polyphosphates in vertebrate tissues. J Neurochem 18:1229–1242

Gogvadze V, Orrenius S, Zhivotovsky B (2009) Mitochondria as targets for chemotherapy. Apoptosis 14:624–640

Gray MJ, Wholey WY, Wagner NO, Cremers CM, Mueller-Schickert A, Hock NT, Jakob U et al (2014) Polyphosphate is a primordial chaperone. Mol Cell 53:689–699

Guitart-Mampel M, Urquiza P, Carnevale Neto F, Anderson JR, Hambardikar V, Scoma ER, Merrihew GE, Wang L, MacCoss MJ, Raftery D, Peffers MJ, Solesio ME (2022) Mitochondrial inorganic polyphosphate (polyP) is a potent regulator of mammalian bioenergetics in SH-SY5Y cells: a proteomics and metabolomics study. Front Cell Dev Biol 10:833127

Guo XZ, Shan C, Hou YF, Zhu G, Tao B, Sun LH, Liu JM et al (2018) Osteocalcin ameliorates motor dysfunction in a 6-hydroxydopamine-induced Parkinson's disease rat model through AKT/GSK3beta signaling. Front Mol Neurosci 11:343

Hacchou Y, Uematsu T, Ueda O, Usui Y, Uematsu S, Takahashi M, Furusawa K et al (2007) Inorganic polyphosphate: a possible stimulant of bone formation. J Dent Res 86:893–897

Hambardikar V, Guitart-Mampel M, Scoma ER, Urquiza P, Nagana GGA, Raftery D, Collins JA, Solesio ME (2022) Enzymatic depletion of mitochondrial inorganic polyphosphate (polyP) increases the generation of reactive oxygen species (ROS) and the activity of the pentose phosphate pathway (PPP) in mammalian cells. Antioxidants (Basel) 11:685

Harada K, Itoh H, Kawazoe Y, Miyazaki S, Doi K, Kubo T, Shiba T et al (2013) Polyphosphate-mediated inhibition of tartrate-resistant acid phosphatase and suppression of bone resorption of osteoclasts. PLoS One 8:e78612

Hassanian SM, Avan A, Ardeshirylajimi A (2017) Inorganic polyphosphate: a key modulator of inflammation. J Thromb Haemost 15:213–218

Hely MA, Reid WG, Adena MA, Halliday GM, Morris JG (2008) The Sydney multicenter study of Parkinson's disease: the inevitability of dementia at 20 years. Mov Disord 23:837–844

Hernandez-Ruiz L, Gonzalez-Garcia I, Castro C, Brieva JA, Ruiz FA (2006) Inorganic polyphosphate and specific induction of apoptosis in human plasma cells. Haematologica 91: 1180–1186

Hofmann AD, Beyer M, Krause-Buchholz U, Wobus M, Bornhauser M, Rodel G (2012) OXPHOS supercomplexes as a hallmark of the mitochondrial phenotype of adipogenic differentiated human MSCs. PLoS One 7:e35160

Holmsen H (1985) Platelet metabolism and activation. Semin Hematol 22:219–240

Holmstrom KM, Marina N, Baev AY, Wood NW, Gourine AV, Abramov AY (2013) Signalling properties of inorganic polyphosphate in the mammalian brain. Nat Commun 4:1362

Hsu YC, Wu YT, Yu TH, Wei YH (2016) Mitochondria in mesenchymal stem cell biology and cell therapy: from cellular differentiation to mitochondrial transfer. Semin Cell Dev Biol 52:119–131

Ishige K, Zhang H, Kornberg A (2002) Polyphosphate kinase (PPK2), a potent, polyphosphate-driven generator of GTP. Proc Natl Acad Sci U S A 99:16684–16688

Jahid IK, Silva AJ, Benitez JA (2006) Polyphosphate stores enhance the ability of *Vibrio cholerae* to overcome environmental stresses in a low-phosphate environment. Appl Environ Microbiol 72:7043–7049

Judge S, Leeuwenburgh C (2007) Cardiac mitochondrial bioenergetics, oxidative stress, and aging. Am J Physiol Cell Physiol 292:C1983–C1992

Kawazoe Y, Shiba T, Nakamura R, Mizuno A, Tsutsumi K, Uematsu T, Kohgo T et al (2004) Induction of calcification in MC3T3-E1 cells by inorganic polyphosphate. J Dent Res 83:613–618

Kawazoe Y, Katoh S, Onodera Y, Kohgo T, Shindoh M, Shiba T (2008) Activation of the FGF signaling pathway and subsequent induction of mesenchymal stem cell differentiation by inorganic polyphosphate. Int J Biol Sci 4:37–47

Khong ML, Li L, Solesio ME, Pavlov EV, Tanner JA (2020) Inorganic polyphosphate controls cyclophilin B-mediated collagen folding in osteoblast-like cells. FEBS J 287:4500–4524

Khrimian L, Obri A, Karsenty G (2017) Modulation of cognition and anxiety-like behavior by bone remodeling. Mol Metab 6:1610–1615

Kilarski LL, Pearson JP, Newsway V, Majounie E, Knipe MD, Misbahuddin A, Morris HR et al (2012) Systematic review and UK-based study of PARK2 (parkin), PINK1, PARK7 (DJ-1) and LRRK2 in early-onset Parkinson's disease. Mov Disord 27:1522–1529

Kocer A, Yaman A, Niftaliyev E, Duruyen H, Eryilmaz M, Kocer E (2013) Assessment of platelet indices in patients with neurodegenerative diseases: mean platelet volume was increased in patients with Parkinson's disease. Curr Gerontol Geriatr Res 2013:986254

Kornberg A (1995) Inorganic polyphosphate: toward making a forgotten polymer unforgettable. J Bacteriol 177:491–496

Kornberg A, Rao NN, Ault-Riche D (1999) Inorganic polyphosphate: a molecule of many functions. Annu Rev Biochem 68:89–125

Koupenova M, Clancy L, Corkrey HA, Freedman JE (2018) Circulating platelets as mediators of immunity, inflammation, and thrombosis. Circ Res 122:337–351

Kumble KD, Kornberg A (1995) Inorganic polyphosphate in mammalian cells and tissues. J Biol Chem 270:5818–5822

Kumble KD, Kornberg A (1996) Endopolyphosphatases for long chain inorganic polyphosphate in yeast and mammals. J Biol Chem 271:27146–27151

Lempart J, Jakob U (2019) Role of polyphosphate in amyloidogenic processes. Cold Spring Harb Perspect Biol 11(5). https://doi.org/10.1101/cshperspect.a034041

Lempart J, Tse E, Lauer JA, Ivanova MI, Sutter A, Yoo N, Jakob U et al (2019) Mechanistic insights into the protective roles of polyphosphate against amyloid cytotoxicity. Life Sci Alliance 2(5). https://doi.org/10.26508/lsa.201900486

Li Q, Gao Z, Chen Y, Guan MX (2017) The role of mitochondria in osteogenic, adipogenic and chondrogenic differentiation of mesenchymal stem cells. Protein Cell 8:439–445

Liguori I, Russo G, Curcio F, Bulli G, Aran L, Della-Morte D, Abete P et al (2018) Oxidative stress, aging, and diseases. Clin Interv Aging 13:757–772

Liu T, Zhang L, Joo D, Sun SC (2017) NF-kappaB signaling in inflammation. Signal Transduct Target Ther 2. https://doi.org/10.1038/sigtrans.2017.23

Lotz M, Loeser RF (2012) Effects of aging on articular cartilage homeostasis. Bone 51:241–248

Maas C, Renne T (2018) Coagulation factor XII in thrombosis and inflammation. Blood 131:1903–1909

Maetzler W, Berg D (2018) Parkinson disease in 2017: changing views after 200 years of Parkinson disease. Nat Rev Neurol 14:70–72

Mari D, Ogliari G, Castaldi D, Vitale G, Bollini EM, Lio D (2008) Hemostasis and ageing. Immun Ageing 5:12

Mattson MP (2000) Apoptosis in neurodegenerative disorders. Nat Rev Mol Cell Biol 1:120–129

McColl G, Roberts BR, Pukala TL, Kenche VB, Roberts CM, Link CD, Cherny RA et al (2012) Utility of an improved model of amyloid-beta (Abeta(1)(-)(4)(2)) toxicity in *Caenorhabditis elegans* for drug screening for Alzheimer's disease. Mol Neurodegener 7:57

McCormack JG, Halestrap AP, Denton RM (1990) Role of calcium ions in regulation of mammalian intramitochondrial metabolism. Physiol Rev 70:391–425

McIntyre B, Solesio ME (2021) Mitochondrial inorganic polyphosphate (polyP): the missing link of mammalian bioenergetics. Neural Regen Res 16:2227–2228

Melber A, Haynes CM (2018) UPR(mt) regulation and output: a stress response mediated by mitochondrial-nuclear communication. Cell Res 28:281–295

Morrissey JH, Choi SH, Smith SA (2012) Polyphosphate: an ancient molecule that links platelets, coagulation, and inflammation. Blood 119:5972–5979

Mullan A, Quinn JP, McGrath JW (2002) A nonradioactive method for the assay of polyphosphate kinase activity and its application in the study of polyphosphate metabolism in *Burkholderia cepacia*. Anal Biochem 308:294–299

Müller WEG, Wang XH, Diehl-Seifert B, Kropf K, Schlossmacher U, Lieberwirth I, Schröder HC et al (2011) Inorganic polymeric phosphate/polyphosphate as an inducer of alkaline phosphatase and a modulator of intracellular Ca^{2+} level in osteoblasts (SaOS-2 cells) in vitro. Acta Biomater 7:2661–2671

Müller WEG, Wang SF, Ackermann M, Neufurth M, Steffen R, Mecja E, Wang XH et al (2017a) Rebalancing beta-amyloid-induced decrease of ATP level by amorphous nano/micro polyphosphate: suppression of the neurotoxic effect of amyloid beta-protein fragment 25-35. Int J Mol Sci 18(10). https://doi.org/10.3390/ijms18102154

Müller WEG, Wang SF, Neufurth M, Kokkinopoulou M, Feng QL, Schröder HC, Wang XH (2017b) Polyphosphate as a donor of high-energy phosphate for the synthesis of ADP and ATP. J Cell Sci 130:2747–2756

Müller WEG, Schröder HC, Wang XH (2019) Inorganic polyphosphates as storage for and generator of metabolic energy in the extracellular matrix. Chem Rev 119:12337–12374

Murphy MP, LeVine H 3rd (2010) Alzheimer's disease and the amyloid-beta peptide. J Alzheimers Dis 19:311–323

Nicholls DG (1978) The regulation of extramitochondrial free calcium ion concentration by rat liver mitochondria. Biochem J 176:463–474

Nikel PI, Chavarria M, Martinez-Garcia E, Taylor AC, de Lorenzo V (2013) Accumulation of inorganic polyphosphate enables stress endurance and catalytic vigour in *Pseudomonas putida* KT2440. Microb Cell Factories 12:50

Nuytemans K, Theuns J, Cruts M, Van Broeckhoven C (2010) Genetic etiology of Parkinson disease associated with mutations in the *SNCA*, *PARK2*, *PINK1*, *PARK7*, and *LRRK2* genes: a mutation update. Hum Mutat 31:763–780

Oberle GP, Stahl RA (1990) The acute side effects of nonsteroidal antiphlogistics on the kidneys. Dtsch Med Wochenschr 115:309–314

Ohtomo R, Sekiguchi Y, Kojima T, Saito M (2008) Different chain length specificity among three polyphosphate quantification methods. Anal Biochem 383:210–216

Oikonomou E, Leopoulou M, Theofilis P, Antonopoulos AS, Siasos G, Latsios G, Tousoulis D et al (2020) A link between inflammation and thrombosis in atherosclerotic cardiovascular diseases: clinical and therapeutic implications. Atherosclerosis 309:16–26

Omelon S, Georgiou J, Henneman ZJ, Wise LM, Sukhu B, Hunt T, Grynpas MD et al (2009) Control of vertebrate skeletal mineralization by polyphosphates. PLoS One 4:e5634

Pan X, Kaminga AC, Wen SW, Wu X, Acheampong K, Liu A (2019) Dopamine and dopamine receptors in Alzheimer's disease: a systematic review and network meta-analysis. Front Aging Neurosci 11:175

Paradies G, Petrosillo G, Paradies V, Ruggiero FM (2010) Oxidative stress, mitochondrial bioenergetics, and cardiolipin in aging. Free Radic Biol Med 48:1286–1295

Pavlov E, Aschar-Sobbi R, Campanella M, Turner RJ, Gomez-Garcia MR, Abramov AY (2010) Inorganic polyphosphate and energy metabolism in mammalian cells. J Biol Chem 285:9420–9428

Perier C, Vila M (2012) Mitochondrial biology and Parkinson's disease. Cold Spring Harb Perspect Med 2:a009332

Poirier Y, Grimm A, Schmitt K, Eckert A (2019) Link between the unfolded protein response and dysregulation of mitochondrial bioenergetics in Alzheimer's disease. Cell Mol Life Sci 76: 1419–1431

Priault M, Salin B, Schaeffer J, Vallette FM, di Rago JP, Martinou JC (2005) Impairing the bioenergetic status and the biogenesis of mitochondria triggers mitophagy in yeast. Cell Death Differ 12:1613–1621

Rao NN, Kornberg A (1996) Inorganic polyphosphate supports resistance and survival of stationary-phase *Escherichia coli*. J Bacteriol 178:1394–1400

Rao NN, Kornberg A (1999) Inorganic polyphosphate regulates responses of *Escherichia coli* to nutritional stringencies, environmental stresses and survival in the stationary phase. Prog Mol Subcell Biol 23:183–195

Rao NN, Gomez-Garcia MR, Kornberg A (2009) Inorganic polyphosphate: essential for growth and survival. Annu Rev Biochem 78:605–647

Requejo-Aguilar R, Bolanos JP (2016) Mitochondrial control of cell bioenergetics in Parkinson's disease. Free Radic Biol Med 100:123–137

Ruan L, Zhou C, Jin E, Kucharavy A, Zhang Y, Wen Z, Li R et al (2017) Cytosolic proteostasis through importing of misfolded proteins into mitochondria. Nature 543:443–446

Salomon AR, Voehringer DW, Herzenberg LA, Khosla C (2000) Understanding and exploiting the mechanistic basis for selectivity of polyketide inhibitors of F_0F_1-ATPase. Proc Natl Acad Sci U S A 97:14766–14771

Sandor C, Honti F, Haerty W, Szewczyk-Krolikowski K, Tomlinson P, Evetts S, Wade-Martins R et al (2017) Whole-exome sequencing of 228 patients with sporadic Parkinson's disease. Sci Rep 7:41188

Santini E, Heiman M, Greengard P, Valjent E, Fisone G (2009) Inhibition of mTOR signaling in Parkinson's disease prevents L-DOPA-induced dyskinesia. Sci Signal 2(ra36)

Schröder HC, Kurz L, Müller WEG, Lorenz B (2000) Polyphosphate in bone. Biochemistry (Mosc) 65:296–303

Schulz AM, Haynes CM (2015) UPR(mt)-mediated cytoprotection and organismal aging. Biochim Biophys Acta 1847:1448–1456

Seidlmayer LK, Blatter LA, Pavlov E, Dedkova EN (2012a) Inorganic polyphosphate—an unusual suspect of the mitochondrial permeability transition mystery. Channels (Austin) 6:463–467

Seidlmayer LK, Gomez-Garcia MR, Blatter LA, Pavlov E, Dedkova EN (2012b) Inorganic polyphosphate is a potent activator of the mitochondrial permeability transition pore in cardiac myocytes. J Gen Physiol 139:321–331

Seidlmayer LK, Juettner VV, Kettlewell S, Pavlov EV, Blatter LA, Dedkova EN (2015) Distinct mPTP activation mechanisms in ischaemia-reperfusion: contributions of Ca^{2+}, ROS, pH, and inorganic polyphosphate. Cardiovasc Res 106:237–248

Seidlmayer LK, Gomez-Garcia MR, Shiba T, Porter GA Jr, Pavlov EV, Bers DM, Dedkova EN (2019) Dual role of inorganic polyphosphate in cardiac myocytes: the importance of polyP chain length for energy metabolism and mPTP activation. Arch Biochem Biophys 662:177–189

Shan C, Ghosh A, Guo XZ, Wang SM, Hou YF, Li ST, Liu JM (2019) Roles for osteocalcin in brain signalling: implications in cognition- and motor-related disorders. Mol Brain 12:23

Sinauridze EI, Kireev DA, Popenko NY, Pichugin AV, Panteleev MA, Krymskaya OV, Ataullakhanov FI (2007) Platelet microparticle membranes have 50- to 100-fold higher specific procoagulant activity than activated platelets. Thromb Haemost 97:425–434

Smith SA, Morrissey JH (2014) Polyphosphate: a new player in the field of hemostasis. Curr Opin Hematol 21:388–394

Solesio ME, Pavlov EV (2016) Methods of inorganic polyphosphate (PolyP) assay in higher eukaryotic cells. In: Kulakovskaya T, Pavlov E, Dedkova E (eds) Inorganic polyphosphates in eukaryotic cells. Springer, Cham. https://doi.org/10.1007/978-3-319-41073-9_6

Solesio ME, Demirkhanyan L, Zakharian E, Pavlov EV (2016a) Contribution of inorganic polyphosphate towards regulation of mitochondrial free calcium. Biochim Biophys Acta 1860:1317–1325

Solesio ME, Elustondo PA, Zakharian E, Pavlov EV (2016b) Inorganic polyphosphate (polyP) as an activator and structural component of the mitochondrial permeability transition pore. Biochem Soc Trans 44:7–12

Solesio ME, Garcia Del Molino LC, Elustondo PA, Diao C, Chang JC, Pavlov EV (2020) Inorganic polyphosphate is required for sustained free mitochondrial calcium elevation, following calcium uptake. Cell Calcium 86:102127

Solesio ME, Xie L, McIntyre B, Ellenberger M, Mitaishvili E, Bhadra-Lobo S, Pavlov EV et al (2021) Depletion of mitochondrial inorganic polyphosphate (polyP) in mammalian cells causes metabolic shift from oxidative phosphorylation to glycolysis. Biochem J 478:1631–1646

Sonntag KC, Ryu WI, Amirault KM, Healy RA, Siegel AJ, McPhie DL, Cohen BM et al (2017) Late-onset Alzheimer's disease is associated with inherent changes in bioenergetics profiles. Sci Rep 7:14038

Spillantini MG, Schmidt ML, Lee VM, Trojanowski JQ, Jakes R, Goedert M (1997) Alpha-synuclein in Lewy bodies. Nature 388:839–840

Spinelli JB, Haigis MC (2018) The multifaceted contributions of mitochondria to cellular metabolism. Nat Cell Biol 20:745–754

St-Pierre JP, Wang Q, Li SQ, Pilliar RM, Kandel RA (2012) Inorganic polyphosphate stimulates cartilage tissue formation. Tissue Eng Part A 18:1282–1292

Streichan M, Golecki JR, Schön G (1990) Polyphosphate-accumulating bacteria from sewage plants with different proceses for biological phosphorus removal. FEMS Microbiol Ecol 6: 113–124

Suess PM, Watson J, Chen W, Gomer RH (2017) Extracellular polyphosphate signals through Ras and Akt to prime *Dictyostelium discoideum* cells for development. J Cell Sci 130:2394–2404

Swerdlow RH, Parks JK, Cassarino DS, Maguire DJ, Maguire RS, Bennett JP Jr, Parker WD Jr et al (1997) Cybrids in Alzheimer's disease: a cellular model of the disease? Neurology 49:918–925

Szewczyk Z, Zarzycki J, Szepietowski T, Szroeder J, Rak J (1978) Electron microscopic studies of various cases of glomerulonephritis (preliminary report). Patol Pol 29:359–367

Tahara EB, Navarete FD, Kowaltowski AJ (2009) Tissue-, substrate-, and site-specific characteristics of mitochondrial reactive oxygen species generation. Free Radic Biol Med 46:1283–1297

Travers RJ, Smith SA, Morrissey JH (2015) Polyphosphate, platelets, and coagulation. Int J Lab Hematol 37(Suppl 1):31–35

Uccelli A, Moretta L, Pistoia V (2008) Mesenchymal stem cells in health and disease. Nat Rev Immunol 8:726–736

UN (2019) World population ageing 2019 highlights; United Nations, Department of Economic and Social Affairs. New York, NY, USA, Population Division

van der Meijden PEJ, Heemskerk JWM (2019) Platelet biology and functions: new concepts and clinical perspectives. Nat Rev Cardiol 16:166–179

Vila M, Jackson-Lewis V, Guegan C, Wu DC, Teismann P, Choi DK, Przedborski S et al (2001) The role of glial cells in Parkinson's disease. Curr Opin Neurol 14:483–489

Wang C, Youle RJ (2009) The role of mitochondria in apoptosis. Annu Rev Genet 43:95–118

Wang L, Fraley CD, Faridi J, Kornberg A, Roth RA (2003) Inorganic polyphosphate stimulates mammalian TOR, a kinase involved in the proliferation of mammary cancer cells. Proc Natl Acad Sci U S A 100:11249–11254

Wang X, Wang W, Li L, Perry G, Lee HG, Zhu X (2014) Oxidative stress and mitochondrial dysfunction in Alzheimer's disease. Biochim Biophys Acta 1842:1240–1247

Wang Y, Li M, Li P, Teng H, Fan D, Du W, Guo Z (2019) Progress and applications of polyphosphate in bone and cartilage regeneration. Biomed Res Int 2019:5141204

Webster KA (2012) Mitochondrial membrane permeabilization and cell death during myocardial infarction: roles of calcium and reactive oxygen species. Futur Cardiol 8:863–884

Wei J, Shimazu J, Makinistoglu MP, Maurizi A, Kajimura D, Zong H, Karsenty G et al (2015) Glucose uptake and Runx2 synergize to orchestrate osteoblast differentiation and bone formation. Cell 161:1576–1591

Westermann B (2012) Bioenergetic role of mitochondrial fusion and fission. Biochim Biophys Acta 1817:1833–1838

Wong RS (2011) Apoptosis in cancer: from pathogenesis to treatment. J Exp Clin Cancer Res 30:87

Yoo NG, Dogra S, Meinen BA, Tse E, Haefliger J, Southworth DR, Jakob U et al (2018) Polyphosphate stabilizes protein unfolding intermediates as soluble amyloid-like oligomers. J Mol Biol 430:4195–4208

Zorov DB, Juhaszova M, Sollott SJ (2014) Mitochondrial reactive oxygen species (ROS) and ROS-induced ROS release. Physiol Rev 94:909–950

Zurawska-Plaksej E, Kuliczkowski W, Karolko B, Cielecka-Prynda M, Debski J, Kaaz K, Usnarska-Zubkiewicz L et al (2020) Platelet polyphosphate level is elevated in patients with chronic primary thrombocytopenia: a preliminary study. Adv Clin Exp Med 29:1051–1056

Chapter 4
Polyphosphate in Chronic Wound Healing: Restoration of Impaired Metabolic Energy State

Xiaohong Wang, Hadrian Schepler, Meik Neufurth, Shunfeng Wang, Heinz C. Schröder, and Werner E. G. Müller

Abstract Many pathological conditions are characterized by a deficiency of metabolic energy. A prominent example is nonhealing or difficult-to-heal chronic wounds. Because of their unique ability to serve as a source of metabolic energy, inorganic polyphosphates (polyP) offer the opportunity to develop novel strategies to treat such wounds. The basis is the generation of ATP from the polymer through the joint action of two extracellular or plasma membrane-bound enzymes alkaline phosphatase and adenylate kinase, which enable the transfer of energy-rich phosphate from polyP to AMP with the formation of ADP and finally ATP. Building on these findings, it was possible to develop novel regeneratively active materials for wound therapy, which have already been successfully evaluated in first studies on patients.

Keywords Inorganic polyphosphate · Wound healing · Nanoparticles · Compressed collagen · Extracellular matrix · Zinc ions · Calcium ions · Cell migration · Angiogenesis · Human epidermal keratinocytes

---

Xiaohong Wang and Hadrian Schepler contributed equally to this work.

X. Wang · M. Neufurth · S. Wang · H. C. Schröder · W. E. G. Müller (✉)
ERC Advanced Investigator Group, Institute for Physiological Chemistry, University Medical Center of the Johannes Gutenberg University, Mainz, Germany
e-mail: wang013@uni-mainz.de; hschroed@uni-mainz.de; wmueller@uni-mainz.de

H. Schepler
Department of Dermatology, University Clinic Mainz, Mainz, Germany
e-mail: hadrian.schepler@unimedizin-mainz.de

W. E. G. Müller et al. (eds.), *Inorganic Polyphosphates*, Progress in Molecular and Subcellular Biology 61, https://doi.org/10.1007/978-3-031-01237-2_4

4.1 Introduction

It has been recognized for a long time that wound regeneration is an energy-dependent process (Im and Hoopes 1970). Based on this background, a number of therapies have been developed, with more or less success, which aim to increase the mitochondrial production of ATP. These therapies include, among others, the use of hyperbaric oxygen to upregulate the rate of oxidative phosphorylation (Guo and DiPietro 2010; Han and Ceilley 2017) or the use of photobiomodulation (or low-level laser therapy) based on the absorption of photon (light) energy to rise the mitochondrial membrane potential and thereby ATP generation via the mitochondrial F_1F_O-ATP synthase (Karu 2010; Leyane et al. 2021). All these studies assume that the ATP thereby produced is primarily used for ATP-consuming processes within the cells during wound healing (Kotwal et al. 2015). The ATP demand of extracellular processes or even an extracellular formation of the ATP required has been little, if at all, considered. It was assumed that the comparatively extremely low concentrations of extracellular ATP (~10^{-9} to 10^{-8} M; Trautmann 2009) are mainly of intracellular origin and are required mostly for signaling processes that are initiated by binding of the nucleotide or its hydrolysis products to receptors on the plasma membrane (Yegutkin 2008).

But there are a number of processes taking place in the extracellular space that need a continuous supply of metabolic energy in the form of ATP and whose energy requirements dramatically increase during repair/regeneration processes such as wound healing. These ATP-consuming processes include, among others, the organization and maintenance of the function of the extracellular matrix (ECM) as well as extracellular kinase reactions, *cis-trans* isomerase reactions, and the function of extracellular chaperone(−like) proteins such as clusterin (Müller et al. 2017d, 2019a).

The discovery that inorganic polyphosphate (polyP), which is released from cells such as the blood platelets after activation at the sites of tissue damage, can locally serve for the extracellular production of large amounts of ATP was groundbreaking in the search for the energy source required for these processes (Müller et al. 2015b). We were able to show that alkaline phosphatase (ALP), a ubiquitous enzyme that occurs both intra- and extracellularly (membrane-associated and soluble), is not only able to hydrolyze polyP to orthophosphate but also in interaction with another enzyme, adenylate kinase (ADK), which is also linked to the plasma membrane, transfers phosphate to AMP, which is thereby up-phosphorylated to ADP and ATP (Müller et al. 2019a). The polyP required for this process is particularly enriched in the platelets, in their numerous acidocalcisomes (dense granules) (Docampo et al. 2010; Müller et al. 2015b), but also in other cells and tissues, such as in the skin (Simbulan-Rosenthal et al. 2015).

Chronic wounds or nonhealing wounds (usually defined as wounds that fail to heal after a period of 3 months) represent a significant burden not only for the patients who suffer from these wounds, but also for the healthcare system as a whole. A variety of pathological and physiological factors can underlie the development of chronic wounds, such as circulatory disorders, metabolic disorders (e.g., diabetes), prolonged pressure on the skin tissue, immune deficiency, tumor diseases, infections, or increasing age (Han and Ceilley 2017). These conditions are often associated with impaired tissue oxygenation, leading to hypoxic cell damage and dysfunction. Exogenously administered polyP, which—with its multiple energy-rich phosphoanhydride bonds—can deliver a multifold of the energy present in ATP, offers a possible solution for the therapy of such wounds (Müller et al. 2019a). PolyP can be given either in a soluble form, as Na-polyP, which is readily soluble in water, or in a particulate form, as amorphous polyP nanoparticles (e.g., Ca-polyP nanoparticles or Zn-polyP nanoparticles), which is more resistant to hydrolysis by ALP and can serve as a depot of the polymer. PolyP molecules of specific ranges of chain lengths can be prepared chemically in large quantities and are already widely used, e.g., as food additives. PolyP also has morphogenetic activity (Wang et al. 2016a, c, d). In this chapter, the present state-of-the-art and examples for the application of the energy-rich polyP in wound healing materials are described.

4.2 Wound Healing and Disturbances Leading to Chronic Wounds

The skin is an organ that is subject to constant regeneration and therefore requires large amounts of metabolic energy (Rolfe and Brown 1997; Reinke and Sorg 2012). This organ, the largest in the human body, consists of three layers, hypodermis, dermis, and epidermis (Fig. 4.1). The outer layer, the epidermis, protects the skin from mechanical injury. It contains the keratinocytes, which originate from stem cells located in the basal layer (stratum basale) at the basal lamina that separates the epidermis from the dermis, and migrate outward during their differentiation, via the stratum spinosum, stratum granulosum, stratum lucidum, to the outermost layer, the stratum corneum, where they are finally shed (Yokouchi et al. 2016); Fig. 4.1. The dermis, below the epidermis, gives the skin flexibility. In addition to collagenous, reticular, and elastic fibers, this layer, which is traversed by blood vessels and lymphatics, contains primarily fibroblasts in addition to macrophages, T- and B-lymphocytes, mast cells, and dendritic cells. The hypodermis mainly consists of fat cells and loose connective tissue.

When the epidermis and the underlying dermis are injured, the blood vessels within the dermis rupture, leading to the release of platelets and other blood constituents into the wounded area. Upon activation, the platelets release a number of factors that initiate the blood coagulation cascade. A fibrin clot forms, which seals the injured tissue (Fig. 4.1). In addition to increased platelet aggregation, a release of pro-inflammatory cytokines and growth factors is observed. After this initial hemostasis phase (blood clotting), wound healing goes through a number of other phases that can overlap in time: the inflammatory phase, the granulation phase, and the epithelialization phase (Singer and Clark 1999; Gosain and DiPietro 2004; Guo and

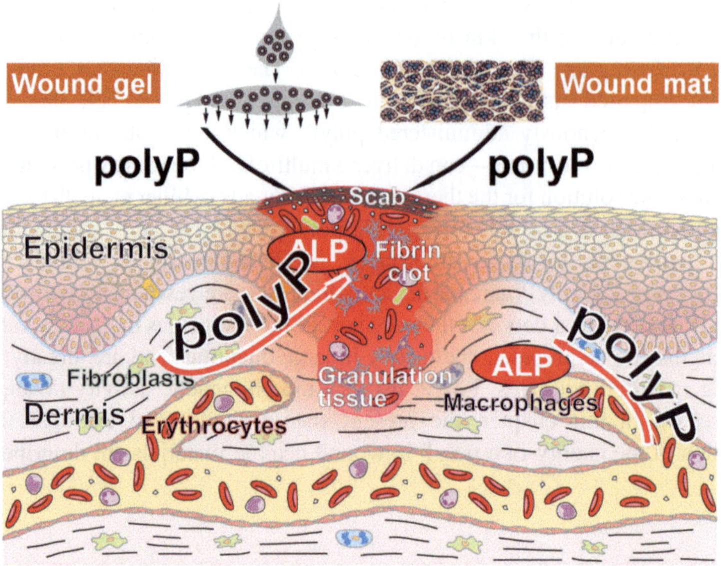

Fig. 4.1 Physiological and therapeutic delivery of polyP to promote wound healing. This sketch shows the structure of the human skin with the epidermis and dermis, which are separated from each other by a basement membrane, the basal lamina. The dermis is crossed by blood vessels and lymph vessels. When injured, platelets emerge, releasing a number of factors that activate the blood clotting cascade. This leads to the formation of a fibrin clot that seals the injured tissue. The factors physiologically released from the platelets also include polyP, which stimulates vascular ingrowth (angiogenesis), proliferation of fibroblasts, and extracellular matrix formation (formation of granulation tissue) as well as reepithelialization. These are metabolic energy (ATP)-consuming processes. The ATP deficiency in chronic wounds can be compensated for by administration of exogenous polyP, in the form of either a polyP-containing wound gel or a polyP-containing wound mat based on compressed collagen

DiPietro 2010; Takeo et al. 2015; Rittié 2016). The inflammatory phase is characterized by an infiltration of inflammatory cells such as lymphocytes, macrophages, and neutrophils. The latter cells produce reactive oxygen species and proteases that exhibit antimicrobial activity against invading microbes. The activity of the macrophages with the release of inflammatory cytokines peaks at the end of the inflammatory phase, before the transition to the regenerative phase (Mosser and Edwards 2008). The latter phase, granulation phase or regenerative phase, is characterized by enhanced cell proliferation and the synthesis of the ECM and ends in the restoration of the epithelium (epithelialization phase).

In chronic wounds, this sequence of wound healing phases is disrupted (Han and Ceilley 2017). The normal production of cytokines and growth factors is impaired (Kiritsi and Nyström 2018; Beidler et al. 2009). Disturbances in angiogenesis lead to an insufficient supply of oxygen and nutrients to the tissue, which leads to chronic hypoxia. Even with normal wound healing, the occurrence of a hypoxic environment can initially be observed (Tandara and Mustoe 2004), but this condition is dramatically increased in chronic wounds. The consequences are severe tissue damage.

Adequate vascular supply of oxygen and nutrients for ATP production is required for the regenerating tissue in the wound area. However, vascularization is impaired with increasing age (Ungvari et al. 2018), with metabolic diseases such as diabetes (Paneni et al. 2013) or ischemia (Kobayashi et al. 2017). The result is a reduced rate of wound healing and ultimately the development of chronic wounds that do not heal. Metabolic energy is also needed for angiogenesis (Barysch and Läuchli 2020). Microvascularization, the initial stage of angiogenesis, depends on cell migration, a process based on chemotaxis. Using HUVEC cells, we were able to show that ATP acts as a chemotactic signal for endothelial cells (Müller et al. 2018a). The extracellular content of ATP can be increased by release of the nucleotide from the cells in response to mechanical stimuli (Moehring et al. 2018) but ATP is also produced extracellularly from polyP, as we have been able to show.

It has long been known that platelets play a crucial role in wound healing (Reinke and Sorg 2012; Nurden 2018). Platelets are rich in polyP, which is secreted by these cells after activation (Morrissey et al. 2012; Müller et al. 2015b, 2019a). The platelet polyP is stored in a condensed, particulate form in their acidocalcisomes (dense granules) (Docampo et al. 2010; Morrissey et al. 2012; Müller et al. 2015b). After activation, the polymer is released from these cells along with a number of other factors. We could show that polyP is a primary source of extracellular ATP production (Müller et al. 2017e). PolyP efficiently promotes the wound healing process including fibroblast proliferation, extracellular matrix formation, and reepithelialization as well as the ingrowth of new blood vessels (angiogenesis) (Schepler et al. 2022).

Using a biomimetic approach, we succeeded in using synthetic polyP to prepare amorphous polyP nanoparticles in the same size range (about 100 nm) and morphology as found in the platelet acidocalcisomes (Müller et al. 2019a). As described below, these polyP nanoparticles, made from metal salts of polyP such as Ca-polyP or Zn-polyP, can be used therapeutically as a depot form in addition to free, soluble polyP (sodium salt; an immediately available form of polyP) in order to treat the injured tissue in the case of an insufficient physiological supply with polyP under certain pathological or physiological conditions. This also makes it possible to treat chronic wounds. The ATP required is formed extracellularly from polyP by the coupled pair of the two enzymes ALP and ADK (Müller et al. 2017e). Both enzymes are present in the wound exudate and in the damaged tissue. PolyP can be used either in the form of a gel or wound mat/dressing containing polyP nanoparticles (depot form) either alone or together with soluble polyP (Na-polyP) (Fig. 4.1).

4.3 PolyP Synthesis, Distribution, and Role in Wound Repair

The energy delivering polyP is a linear polymer of phosphate (P_i) residues that are linked together via high-energy phosphoanhydride bonds (Müller et al. 2019a), Fig. 4.2a. It is the first discovered bioinorganic polymer that provides metabolic energy. Physiologically, this polymer serves both as an energy store and as an energy donor, not only intracellularly but also, in particular, extracellularly.

PolyP is synthesized in the mitochondria within the cells and possibly by a polyP polymerase complex in the acidocalcisomal membrane and stored in these cell organelles (acidocalcisomes), including the platelet dense granules (Fig. 4.2b). The chemical energy stored in the phosphoanhydride bonds of polyP is released during the enzymatic hydrolysis of the polymer by ALP (Lorenz and Schröder 2001), an enzyme that is membrane-bound on the outer side of the plasma membrane but also occurs as a free enzyme both intracellularly and extracellularly (Müller et al. 2019a). There, at the outer cell membrane or within the cell and in the extracellular space, polyP serves as an ATP generator through the interaction of ALP with the likewise membrane-bound or free ADK (Müller et al. 2019a). The ALP-mediated transfer of the high-energy phosphate to AMP produces ADP, which is then interconverted by ADK into AMP and ATP (Fig. 4.2b). Thus, the energy stored in polyP is converted into metabolically usable energy (Müller et al. 2015b, 2017e, 2018a).

In the platelets, which are cleaved off by the megakaryocytes, polyP is distributed throughout the body via the blood circulation (Schröder et al. 2019); Fig. 4.2c. After passing through the lungs and being transported via the blood vessels, polyP is released at the injured tissue sites to serve to heal wounds or repair bone and cartilage defects. The polyP released by the platelets usually has chain lengths of less than 100 units, usually around 40 P_i residues (Morrissey et al. 2012; Müller et al. 2019a). The two enzymes required for ATP generation at the tissue lesions, ALP (catalyzing the phospho-transfer from polyP to AMP to form ADP) and ADK (catalyzing the conversion of ADP to AMP and ATP), work closely together on the damaged site.

In previous studies, we have shown that polyP has the potential to accelerate tissue regeneration/repair of bone and cartilage defects (Müller et al. 2015b, 2016, Neufurth et al. 2017; reviewed in: Wang et al. 2016c, 2018). For these studies, we developed a biomimetic procedure for the fabrication of polyP nanoparticles, mimicking those found in the platelets (Müller et al. 2015f). The synthetic particles turned out to be morphogenetically active, as shown in both in vitro and in vivo experiments (Wang et al. 2016a, c, d). PolyP induces the genes involved in mineralization and cartilage formation, such as the expression of ALP (Müller et al. 2011) and collagen expression (Müller et al. 2015c), as well as the expression of genes involved in the differentiation of the mesenchymal stem cells to bone and cartilage cells (Wang et al. 2016b; Müller et al. 2017c). Subsequent in vivo studies revealed that the polyP nanoparticles also significantly accelerate wound healing both in normal and in diabetic mice showing impaired healing (Müller et al. 2017b). In the presence of the polyP particles, there is an upregulation of collagen gene

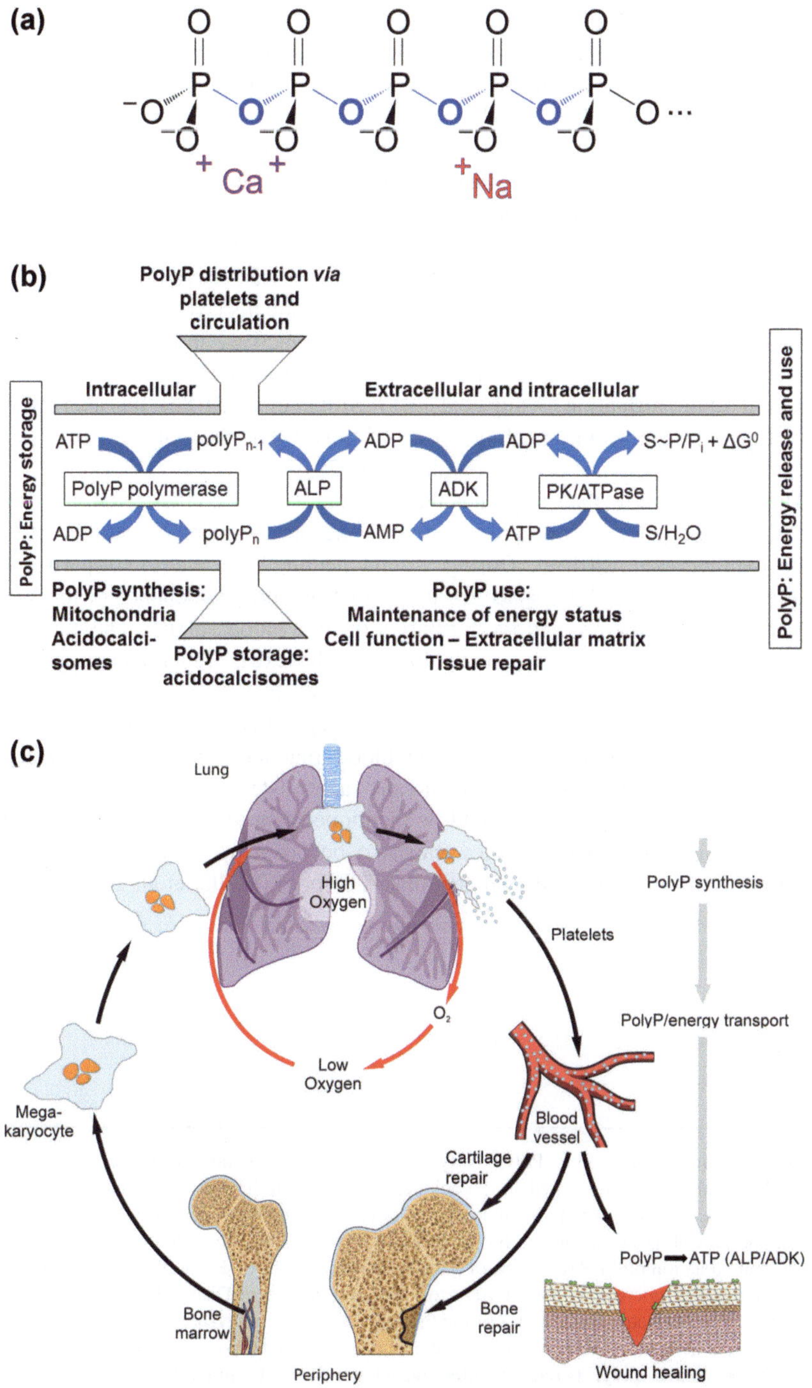

Fig. 4.2 Synthesis, distribution, and use of energy-rich polyP in energy-consuming repair of tissue injuries. (**a**) Structural formula of polyP. Both divalent (e.g., Ca^{2+}) and monovalent (e.g., Na^+) ions can serve as counterions to the polyanionic molecule. (**b**) Energy storage and energy release and use

expression (Müller et al. 2015c) as well as the expression of α-smooth muscle actin and plasminogen activator inhibitor-1 (Müller et al. 2017b). Furthermore, an increase in the ATP level in the wound area and an induction of vascularization are found (Müller et al. 2017b). The direct application of ATP to wounds showed a similarly strong effect on tissue regeneration as polyP (Sarojini et al. 2017). Histological studies revealed a sharp increase in the formation of granulation tissue in wounds treated with the nanoparticles, as well as the development of microvessels within the newly formed connective tissue (Müller et al. 2017b). We were also able to demonstrate the promoting effect of polyP on microvascularization using the tube forming assay with human umbilical vein endothelial cells (HUVEC) (Müller et al. 2018b). The experiments showed that the dramatic increase in the ATP concentration in the cell supernatant after exposure of the cells to polyP is accompanied by an increased tubule formation (Wang et al. 2017; Müller et al. 2018b). It is also relevant with regard to the wound-healing effect of polyP that the polymer also shows antibacterial and antiviral activity (Müller et al. 2017a).

These results show that polyP has the potential to provide the metabolic energy that is lacking, especially in chronic wounds and needed to regenerate the injured tissue (Müller et al. 2015e). During wound healing, this energy is particularly required in the extracellular space for the formation and maintenance of the supramolecular ECM structure and the formation of new blood vessels. The amount of ATP released by the cells is usually low and the extracellular matrix lacks the energy-generating metabolic pathways such as glycolysis and cell organelles such as mitochondria that produce ATP intracellularly. This energy can be provided extracellularly by polyP, either via the blood platelets or, if their capacity is exhausted, by exogenously, therapeutically administered polyP (Müller et al. 2015b; Wang et al. 2016c). It should be mentioned that polyP, when administered in the form of polyP nanoparticles, can also be taken up by cells and can also contribute to an increase in the intracellular ATP pool (Müller et al. 2017f).

Fig. 4.2 (continued) by polyP. PolyP is synthesized in the mitochondria and possibly on the acidocalcisomal membrane, stored in the acidocalcisomes including the acidocalcisomes (dense granules) of the platelets, which distribute the polymer throughout the body via the blood circulation. At the sites of tissue damage, polyP is released from the platelets and used for ATP generation via the mutual action of ALP (catalysis of phosphotransfer from polyP to AMP to form ADP) and ADK (catalysis of interconversion of ADP to AMP and ATP). (**c**) Distribution of polyP via the circulation. The platelets, which split off from the megakaryocytes, serve as carriers of polyP, which is stored in the acidocalcisomes of these cells/precursor cells. After passing the lung and transport via the blood vessels, polyP is released after platelet activation at injured tissue sites and used for wound healing or repair of bone and cartilage. Adapted with permission from ref. Schröder et al. (2019). Copyright 2019, IOP Publishing Ltd.

4.4 PolyP Particles and Coacervate Formation

In recent years, we have succeeded in developing amorphous nanoparticles from different metal salts of polyP (Wang et al. 2018). These nanoparticles have proven to be stable and are converted into a coacervate at their site of action, in situ, after contact with proteins such as protein-containing wound secretions and only then become biologically active and easily biodegradable (Müller et al. 2018c). The amorphous polyP nanoparticles are produced in an alkaline medium, at pH 10, using water-soluble Na-polyP in the presence of a stoichiometric excess of metal ions, e.g., Ca^{2+}, Mg^{2+}, or Zn^{2+} (Müller et al. 2015f). Na-polyP with an average chain length of 30 or 40 phosphate units is mostly used. It is important that the pH of the suspension is kept at pH 10 during the reaction; at lower pH values (pH 7) a coacervate is obtained. The average size of the particles varies between 120 and 150 nm. By varying the manufacturing protocol, it is also possible to prepare smaller or larger particles. On the other hand, the hydrogel-like coacervates which are obtained when the reaction is carried out at neutral pH (pH 7) have a viscous consistency and a high water content. The amorphous state of the polyP nanoparticles is crucial for their biological effectiveness such as their wound-healing activity, since only in this state an easy conversion to the coacervate state is possible, mediated by ionic and other interactions of organic (protein) molecules with the inorganic polymer (Müller et al. 2015f).

In Fig. 4.3, an idealized model of the structure of a Ca-polyP nanoparticle is shown. This model helps to explain the development of the negative zeta potential of the particles after suspension in aqueous solution, which determines the stability of the particles. The model shows that the positively charged calcium ions that are linked via ionic bonds with the negatively charged, polyanionic polyP point outward on the surface of the particles. These calcium ions form the so-called Stern layer (which is, in this example, positively charged). Consequently, a layer of predominantly negatively charged ions from the solution is formed (forming the slip plane), such as hydroxide ions, chloride ions originating from the $CaCl_2$ used for fabrication of the Ca-polyP particles, as well as hydrogen phosphate and dihydrogen phosphate ions as hydrolysis products of the polyP. This outer layer leads to a negative overall charge of the nanoparticles, which thus have a negative zeta potential. Due to their negative zeta potential, the nanoparticles do not tend to aggregate and are stable over long periods of time.

In addition to calcium ions, we used zinc ions as counterions to polyP in the production of polyP nanoparticles (Müller et al. 2018c) due to their anabolic effect on wound healing (Lin et al. 2017). Zinc promotes the epithelization of wounds (Kogan et al. 2017; Lansdown et al. 2007) and is often used in the form of zinc oxide ointments for wound treatment (Lansdown 1993). It is also a cofactor of metalloenzymes involved in tissue repair (Lin et al. 2017). The matrix metalloproteinases belong to these zinc-dependent enzymes (Agren 1993; Mirastschijski et al. 2004). In addition, it has been shown that, in addition to

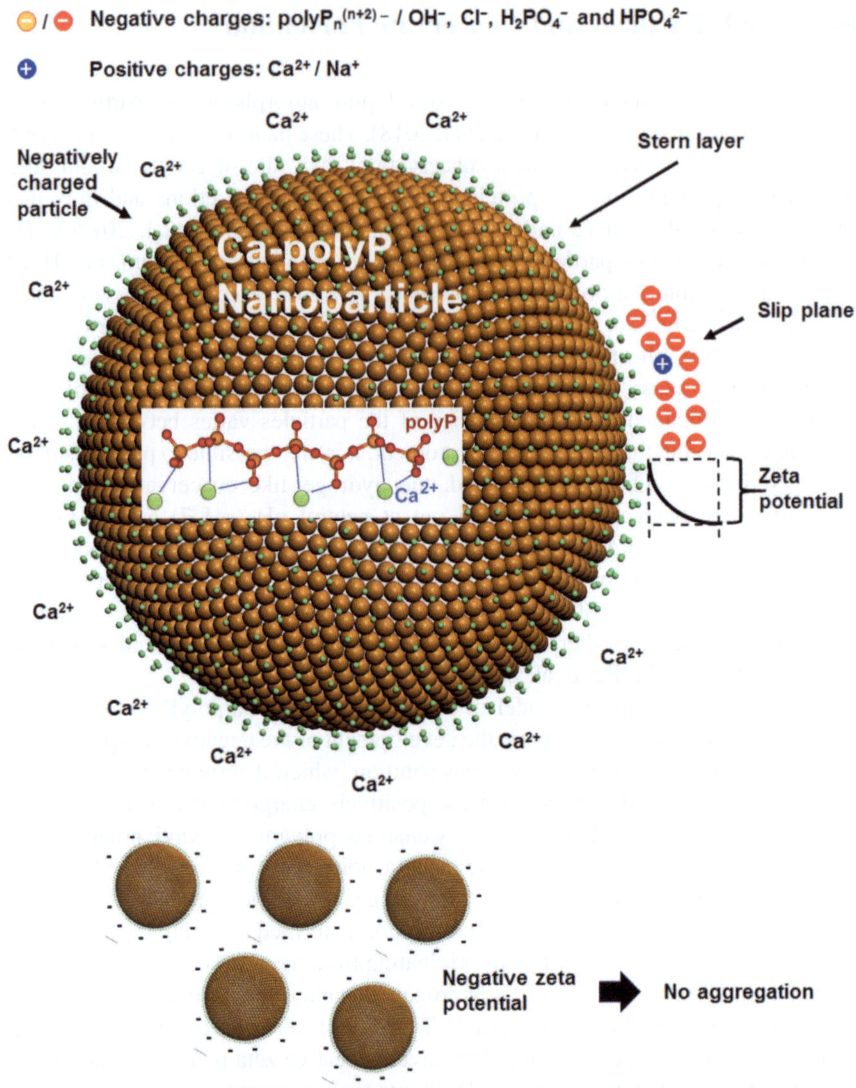

Fig. 4.3 Top. Model of the structure of a Ca-polyP nanoparticle and formation of the negative zeta potential after suspension in aqueous solution. The phosphate units of the polyP chains that carry a negative charge and the positively charged Ca^{2+} ions are shown true-to-scale. The inset in the nanoparticle shows a ball-and-stick model of a section of the Ca-polyP chain. The negatively charged Ca-polyP nanoparticle is covered on its surface with positively charged Ca^{2+} ions, forming the so-called Stern layer. As a result, an outer layer (slip plane), covering the Stern layer, of predominantly negatively charged ions from the solution is formed, consisting of OH^-, Cl^-, $H_2PO_4^-$, and HPO_4^{2-} ions, which results in a negative overall charge of the particle (negative zeta potential). Bottom. Due to their negative zeta potentials, the nanoparticles suspended in the solution do not tend to aggregate. In this form, the particles are stable over long time periods

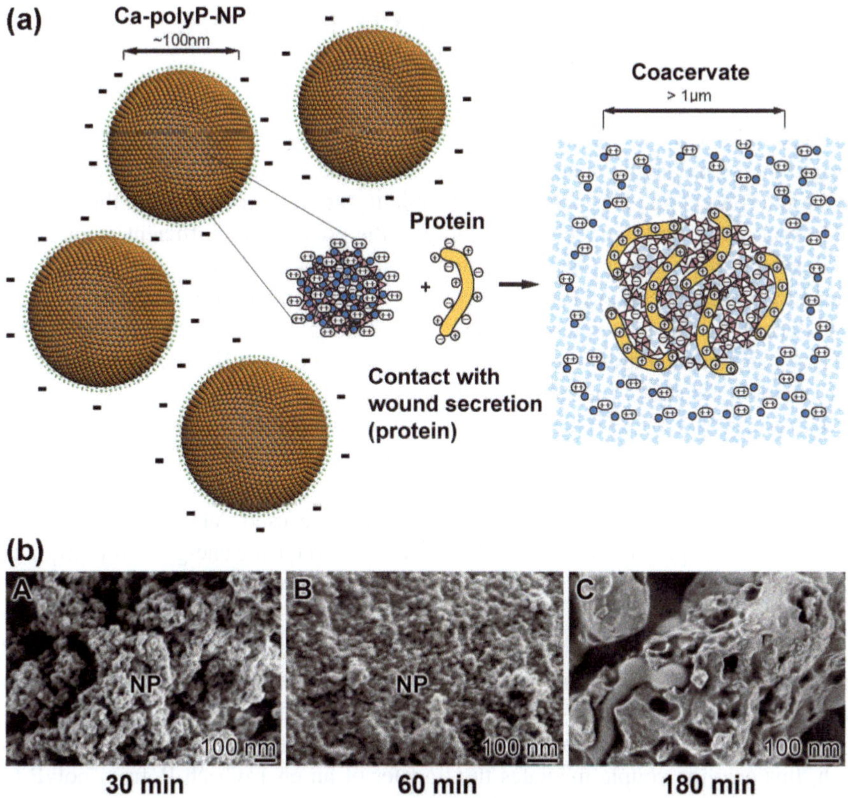

Fig. 4.4 Formation of a coacervate from Ca-polyP nanoparticles after contact with wound exudate. (**a**) In the presence of protein-containing body fluids, such as wound secretions, the Ca-polyP nanoparticles (NP) which have an overall negative charge are broken down and polyP, together with the protein, forms a water-rich coacervate, the physiologically active form of the polymer, by displacing a corresponding amount of Ca^{2+} ions. (**b**) The time kinetics of coacervate formation from Ca-polyP-NP is shown (**b-A** to **b-C**). Adapted with permission from Müller et al. (2019b). Copyright 2019, Portland Press

polyP, zinc can induce an increase in the intracellular ATP level (Pavlica et al. 2009) and promote angiogenesis (Yu et al. 2016).

In the presence of protein-containing body fluids, such as wound exudate, the polyP nanoparticles (Ca-polyP or Zn-polyP nanoparticles), which carry an overall negative charge, are converted into a water-rich coacervate, the physiologically active form of the polymer, whereby the protein, as a result of the positively charged side chains of its basic amino acids, displaces a corresponding amount of metal ions (Fig. 4.4a). The time kinetics of coacervate formation of Ca-polyP nanoparticles is shown in Fig. 4.4b (A-C).

4.5 Sensing ATP Deficiency and ATP Restoration: ALP/ADK Versus AMPK System

As described above, wound healing is a process that strongly depends on the availability of metabolic energy and exogenously administrated polyP, e.g., biomimetic amorphous polyP nanoparticles, is able to provide this energy in the form of ATP, both intra- and extracellularly. Only in the presence of sufficient amounts of ATP, protein biosynthesis and cell proliferation and differentiation, as well as signal transduction can occur. ATP is also required for the induction of angiogenesis, the vascularization of the newly formed regenerating tissue, which involves an ATP-dependent migration of endothelial cells (Müller et al. 2018b). ATP also serves as a chemotactic signal for the cells (Müller et al. 2018a).

The crucial role of ATP in wound healing is also reflected by the observation that the extracellular ATP level increases when repair begins (Yin et al. 2007). Therefore, ATP has been used directly to promote wound healing, by intracellular delivery in wound tissue (Mo et al. 2020). In our approach, we used ATP in a sustainable, storage form, as polymeric polyP, which acts as a donor for energy-rich phosphate for the generation of ATP, both within the cell and in the extracellular space (Müller et al. 2017b, 2020).

Based on our studies, we propose that the extracellular sensor system for the ATP level available, or better to say, the AMP:ATP ratio is formed by the enzyme couple ALP/ADK. The ALP/ADK system could sense the AMP:ATP ratio not only extracellularly but also intracellularly. In case of ATP deficiency, if the AMP:ATP ratio is high, this enzyme couple mediates the transfer of an energy-rich P_i from polyP to AMP to form ADP, catalyzed by ALP, which is then up-phosphorylated to ATP by ADK, which interconverts two ADP to AMP and ATP. If sufficient ATP is available (low AMP:ATP level), ALP will stop the phospho-transfer reaction (ATP generation of the ALP/ADK pair) and only catalyze the hydrolytic breakdown of the polymer, leading to the dissipation of the chemically bound energy of the phosphoanhydride linkages of the polyP as heat.

Such a system acting as an extracellular energy sensor is new. Hitherto, only sensing systems for the intracellular "energy charge" ([ATP] + ½ [ADP] / [ATP] + [ADP] + [AMP]; Atkinson 1968) have been known. In addition, the ALP/ADK system is fast, responding immediately to changes in the extracellular energy level. It markedly differs from the AMP-dependent kinase (AMPK), the second, and most well-known, system detecting ATP deficiency.

The differential responses of the ALP/ADK and AMPK systems are summarized in Fig. 4.5. Both systems differ not only in their location but also in their time kinetics of ATP generation. While the ALP/ADK system senses the extracellular energy state (an intracellular energy-sensing function is also conceivable), the AMPK only acts as an intracellular energy sensor (Hardie 2007). This serine/threonine kinase is a heterotrimer consisting of a catalytic α subunit and two regulatory β and γ subunits (Oakhill et al. 2009). The AMPK is crucially involved in the regulation of intracellular energy homeostasis. The activation of the enzyme

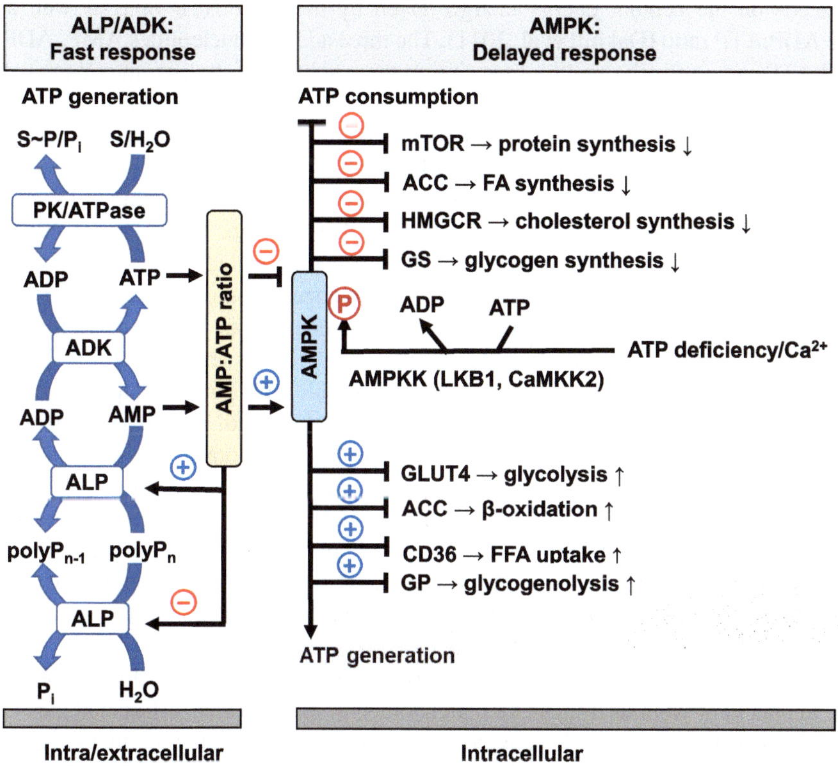

Fig. 4.5 Differential response of the ALP/ADK and AMPK systems to ATP deficiency. Both systems differ not only in their location (ALP/ADK: both intra- and extracellular; AMPK: only intracellular) but also in the time kinetics of ATP regeneration. The ALP/ADK system enables fast, immediate response, at least as long as polyP, the storage form of energy-rich phosphate, is available. The reaction thereby catalyzed depends on the AMP:ATP ratio. If this ratio is high (in case of ATP deficiency), ALP will catalyze the phospho-transfer reaction from polyP to AMP to form ADP that is subsequently interconverted by ADK to ATP and AMP, following the chemical equilibrium. If the AMP:ATP ratio is low (ATP is available), ALP will switch from an enzyme involved in ATP generation (as ALP/ADK couple) to an enzyme that only hydrolyses polyP, thereby dissipating the energy stored in the phosphoanhydride bonds of the polymer in the form of heat. Alike the ALP/ADK system, the AMPK acts as an energy sensor, but only intracellularly. The AMPK only senses the intracellular energy status. AMP activates the enzyme, while ATP antagonizes the AMP effect. The activation of the enzyme depends on phosphorylation of a threonine residue at the α-subunit of AMPK by AMPK kinases (AMPKK), such as liver kinase B1 (LKB1) and calcium/calmodulin-dependent protein kinase kinase 2 (CaMKK2). In contrast to the ALP/ADK system, which leads to an immediate restoration of the ATP pool via the coupled ALP/ADK reaction, ATP regeneration by AMPK involves a series of multistep reactions, resulting in a delayed response and following two different principles: either by reducing ATP consumption or by promoting ATP generation. AMPK-mediated inactivation of ATP-consuming metabolic pathways includes a decrease in protein synthesis by inhibition of mammalian target of rapamycin (mTOR), shutting down fatty acids (FA) synthesis by inactivation of acetyl-CoA carboxylase (ACC), downregulation of cholesterol synthesis by inactivation of 3-hydroxy-3-methyl-glutaryl-CoA reductase (HMGCR), and reduction of glycogen synthesis by inhibition of glycogen synthase (GS). Activation of metabolic pathways resulting in an upregulation of ATP generation include activation of glycolysis by stimulation of glucose uptake by glucose transporter 4 (GLUT4), phosphorylation of acyl CoA-carboxylase (ACC), leading to an inhibition of FA synthesis and

depends on the cellular energy charge, given by the AMP:ATP ratio as well as the ADP:ATP ratio (Oakhill et al. 2011). The three adenine nucleotides AMP, ADP, and ATP competitively bind to three sites in the α subunit of the enzyme (Yan et al. 2018). The AMPK is activated at a low AMP:ATP ratio. It is activated by phosphorylation of a threonine residue of the α-subunit of the enzyme by liver kinase B1 (LKB1) and Ca^{2+}/calmodulin-dependent kinase β (CaMKKβ). The cellular energy balance is restored by AMPK via two mechanisms, both by inhibiting ATP consuming pathways and by promoting pathways that generate ATP (Hardie 2004). ATP consuming pathways inhibited by AMPK activation include the biosynthesis of proteins, fatty acids, and cholesterol, as well as glycogen synthesis. The reduction of protein synthesis is due to an inhibition of the mammalian target of rapamycin complex 1 (mTOR1) caused by AMPK activation (Rodríguez et al. 2021; Trefts and Shaw 2021). The fatty acid and cholesterol biosynthesis are suppressed by AMPK via inactivation of the acetyl-CoA carboxylase and 3-hydroxy-3-methylglutaryl-CoA reductase (Carling et al. 1987). Energy storage via glycogen is suppressed by AMPK through inhibition of glycogen synthase. Metabolic pathways leading to ATP generation after activation of AMPK include an increased glucose uptake via an increased expression and membrane translocation of the glucose transporter 4 (GLUT4), as well as an increased β-oxidation and uptake of free fatty acids via CD36, and stimulation of glycogenolysis by activation of glycogen phosphorylase (Hardie et al. 2006, 2016).

Based on these properties, AMPK has been studied as a possible drug target for the treatment of wounds of diabetic patients such as diabetic ulcers (Lin et al. 2014). Compounds acting as AMPK activators were found to accelerate reepithelialization and mitigate inflammation of such wounds. It has also been proposed that AMPK is involved in regulating vascular homeostasis (Omura et al. 2016). Activation of AMPK in endothelial cells leads to a suppression of the generation of reactive oxygen species (ROS) induced by hyperglycemia. ROS are one of the main factors in the development of endothelial dysfunction in diabetic patients (Tousoulis et al. 2013). Therefore, activation of intracellular AMPK is assumed to play an important role in diabetic wound healing by supporting angiogenesis and delivery of oxygen and nutrients to the injured tissue (Abdel Malik et al. 2017). Also, the hypoxia-inducible factor-1α has been implicated in the AMPK-regulated vascular repair mechanism (Abdel Malik et al. 2017; Huang et al. 2021).

It is worth mentioning that AMPK also plays a crucial role in the regulation of energy-dependent processes in bone and cartilage. Disturbances of these processes counteracted by AMPK include the development of osteoarthritis (Wang et al. 2020a; Yi et al. 2021) or intervertebral disc degeneration (Wang et al. 2021). In MC3T3E1 cells, AMPK activation has been shown to stimulate osteoblast differentiation by inducing *Runx2* expression (Jang et al. 2011).

Fig. 4.5 (continued) promotion of β-oxidation, increase free fatty acids (FFA) uptake via CD36, and stimulation of glycogenolysis by activation of glycogen phosphorylase (GP)

Interestingly, polyP (sodium triphosphate and sodium hexametaphosphate) has been reported to increase the phosphorylation and activation of AMPK (Bae et al. 2016). As a result, stimulation of osteoblastic differentiation of cells in vitro and of bone formation in vivo has been observed.

4.6 Wound Dressings

Wound dressings must meet a variety of requirements. They must be able to maintain adequate wound moisture and prevent desiccation, allow gas exchange, absorb wound exudate, remove necrotic tissue, and also have an antiseptic effect to prevent wound infection (Saghazadeh et al. 2018). In addition to conventional wound mats, electrospun wound mats have been developed more recently (e.g., Müller et al. 2015a), which are particularly distinguished by the fact that they allow excellent air and water vapor permeability but prevent the penetration of liquids and microorganisms (for a review, see Schröder et al. 2017). However, most materials used today, such as collagen, are biologically inert and require additional additives in order to achieve efficient wound healing. Among these additives are substances that show antibacterial properties (Kaushika et al. 2019). To date, however, there are no wound mats available that are capable of supplying the large amounts of metabolic energy required for wound healing. The wound mats described in the following represent a new and promising approach. They are based on the use of synthetic polyP or nanoparticles produced with this polymer, which generates the metabolic energy required in the extracellular space with the help of the two enzymes located on the cell surface and also extracellularly in the wound exudate, ALP and ADK. Physiologically, polyP is released from platelets in addition to cytokines and growth factors (Morrissey et al. 2012). It is interesting to mention that platelet-rich plasma, which contains the polymer in the same chain length range as the polyP we used, has also been used with good success as an additive for wound dressings (Chicharro-Alcántara et al. 2018).

4.7 Wound Healing Mats Containing Energy-Delivering PolyP

Collagen-based wound dressings have gained increasing importance in wound therapy due to their beneficial properties (Chattopadhyay and Raines 2014). Collagen has low immunogenicity and is biodegradable. In addition, it is not cytotoxic and does not cause inflammatory reactions. As a scaffold for wound dressings, collagen promotes cell adhesion and cell growth. A disadvantage of biomaterials based on collagen is poor mechanical strength. One way to increase the mechanical strength of collagen is plastic compression (Martin et al. 2017).

We have succeeded in developing collagen wound healing mats that have the novel property of supplying energy (Müller et al. 2020; Schepler et al. 2022). This was achieved by incorporation of energy-delivering amorphous polyP nanoparticles into collagen tightened by mechanical compression. It is possible either to incorporate prefabricated polyP nanoparticles into the collagen by adding the particles before compression or to generate them in situ during collagen gel formation/fibrillogenesis.

The polyP supplemented collagen mats so produced possess two beneficial properties. Collagen mechanically compressed under suitable conditions has a higher density and thus forms a strong wound-covering layer that acts as a barrier against bacterial infection. On the other hand, these wound mats still allow sufficient cell migration into the wound area by fibroblasts and keratinocytes, which is necessary for tissue regeneration. Second, the incorporated polyP component provides the metabolic energy required for regeneration. Thus, a new principle is introduced into wound therapy, which brings a decisive breakthrough, especially for the treatment of chronic wounds that have not been treatable so far: the delivery of the metabolic energy required for this process (Schepler et al. 2022).

We were initially able to demonstrate the advantageous effect of the polyP particles in animal experiments. Topical application of Ca-polyP nanoparticles to experimental wounds resulted in a significant acceleration of the rate of wound healing in both normal and diabetic mice showing delayed healing (Müller et al. 2017b). This could later also be shown in studies on patients with chronic, therapy-resistant wounds, in which the Ca-polyP-nanoparticles-supplemented compressed collagen wound mats exceeded all expectations and showed a therapeutic effect that had not been achieved with any other material before, leading to the complete healing of these wounds (Schepler et al. 2022); see below.

Mechanical compression is a process during which aqueous fluid is squeezed out of the collagen hydrogel. A sketch showing the reorganization of the collagen fibers during the compression and the incorporation of the polyP particles as well as the coacervate formation of the particulate polyP upon contact with wound exudate is shown in Fig. 4.6. Collagen fibrillogenesis is a pH-dependent process (Fig. 4.6a). Collagen (type I) is soluble at pH 3.5 and forms a triple helix from which the *N*-propeptides and *C*-propeptides are sequentially cleaved off, which leads to the self-organization of the collagen molecules with the formation of collagen fibrils. Increasing the pH to pH 6.5 accelerates the rate of fibrillogenesis by reducing the surface charge. A further increase to pH 8.0 results in fiber formation. The Ca-polyP particles are added to the collagen gel prior to raising the pH. During the increase of the pH, polyP is incorporated into the collagen fiber bundles. Before the mechanical compression, the collagen fibers with the Ca-polyP nanoparticles in the mat are randomly arranged, forming a loose structure (Fig. 4.6b, top). During compression, the fibers orient themselves in the longitudinal direction, which leads to increased density and toughness of the fiber mat (middle). Further, it is shown that upon contact with protein-containing wound exudate, the polyP particles are transformed into the physiologically active, energy delivering coacervate, which serves as the substrate for the ATP generating ALP/ADK system (bottom).

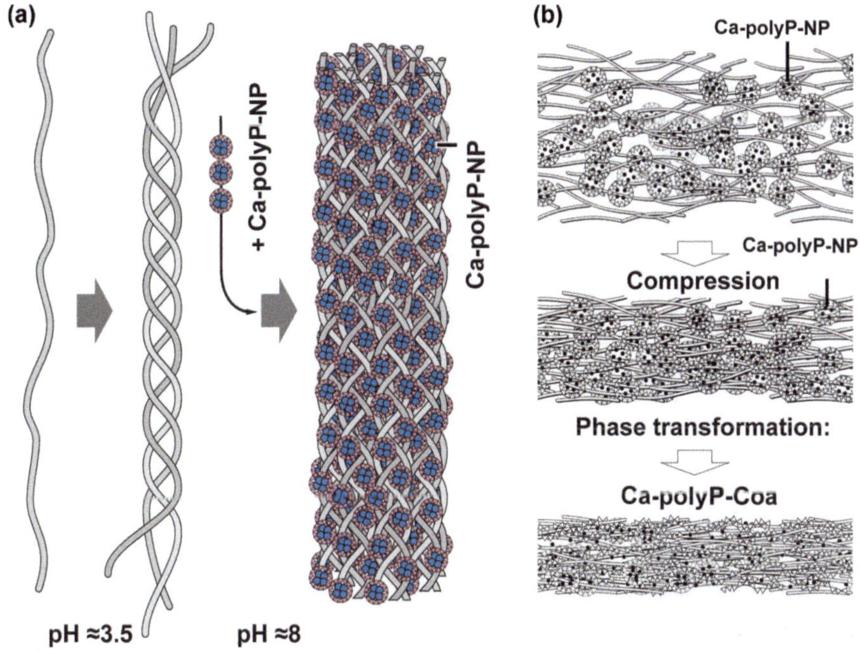

Fig. 4.6 (**a**) Schematic sketch of collagen fibrillogenesis. Tropocollagen, which is soluble at pH 3.5, forms a triple helix (procollagen) by increasing the pH. The procollagen chains subsequently assemble to collagen fibrils/fibers by a further shift of the pH to more alkaline values (pH to ≈8). Ca-polyP nanoparticles become incorporated during this step. (**b**) Scheme showing the reorganization of the collagen fibers during compression and the coacervate formation of the nanoparticle-supplemented collagen mat. The collagen fibers in the mat with the Ca-polyP nanoparticles are randomly arranged before compression (top). During compression, the fibers organize longitudinally, resulting in an enhanced toughness of the mat (middle). Upon contact with protein-containing wound secrete the polyP particles form a Ca-polyP coacervate (Ca-polyP-Coa), the physiologically active form of the polymer (bottom)

The device used to fabricate the compressed collagen mats is shown in Fig. 4.7. In Fig. 4.7a, a schematic drawing of the device is shown, which consists, from top to bottom, of a pressure stamp on which a weight is placed, an upper nylon mesh separating the stamp from the sample, the collagen sample with the Ca-polyP nanoparticles, a perforated Teflon plate in which the stamp, upper nylon mesh, and sample are mounted and placed on a second nylon mesh on a perforated stainless steel sheet and a layer of blotting paper. In Fig. 4.7b, the Teflon plate with the drilled holes fitting to the stamp is shown, into which the collagen solution (collagen type I dissolved in acetic acid) is poured (Fig. 4.7b-A). After self-assembly of the collagen to collagen fibrils induced by shifting the pH from 3.5 to 8.0, the uncompressed collagen sample is covered with the nylon mesh and compression is performed using a weight put on the pressure stamp (Fig. 4.7b-B and b-C).

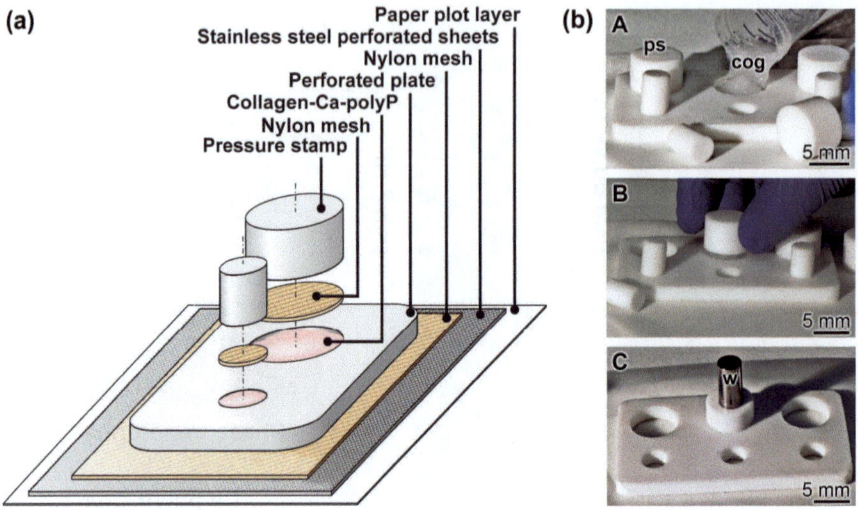

Fig. 4.7 (**a**) Elevation drawing of the device used for manufacturing the compressed collagen mats. The collagen gel is applied on a nylon mesh over a stainless steel perforated sheet and a layer of blotting paper and compressed with the help of a pressure stamp on a second nylon mesh placed on the gel. (**b**) A view of the Teflon-made device with the holes, one of which is filled with collagen gel (cog), the pressure stamp (ps), and the weight (w)

4.8 Cell Growth and Activation on PolyP-Supplemented Collagen Mats

In a first study, compressed collagen mats with incorporated Zn-polyP particles were prepared and their properties examined (Müller et al. 2020). Amorphous Zn-polyP nanoparticles with a size of $\approx$100 nm were used. Like the biomimetic Ca-polyP nanoparticles, Zn-polyP nanoparticles are able to form a physiologically active coacervate after contact with protein and at neutral pH (Müller et al. 2018c).

Human skin keratinocytes growing on these collagen mats showed increased viability and significantly higher metabolic activity compared to keratinocytes on the control mats without the Zn-polyP nanoparticles (Müller et al. 2020). Furthermore, it was found that keratinocytes embedded in the compressed collagen mats spiked with the Zn-polyP particles migrated to the surface of the mats. These cells showed a pronounced development of microvilli, a characteristic of an active state of the cells (Müller et al. 2020). Consequently, an increased migration activity of the keratinocytes was observed (Müller et al. 2020). Well-developed microvilli were not found in keratinocytes cultured in the absence of polyP (Müller et al. 2020).

In order to clarify whether the increased migratory propensity of the keratinocytes is caused by the release of ATP into the extracellular space, experiments were carried out in a Transwell system. The cultures containing the Zn-polyP nanoparticles were co-incubated with apyrase, an enzyme that hydrolyzes ATP. It was found that this leads to complete suppression of cell migration (Müller et al. 2020).

In a further study, Ca-polyP particles were embedded in the compressed collagen mats. These mats were found to greatly improve the activity of keratinocytes and HUVEC cells. The surfaces of the mats supplemented with the Ca-polyP nanoparticles were densely covered with keratinocytes after incubation with the cell cultures, while in the assays with the polyP-free mats, comparatively little attachment of the cells to the mat surface was found (Schepler et al. 2022). Also, the keratinocytes grown on the polyP-enriched mat developed long microvilli and showed increased motility. While keratinocytes plated on the polyP-free collagen mats remained predominantly on the mat surface, keratinocytes seeded on collagen mats supplemented with Ca-polyP nanoparticles migrated deep into the collagen matrix (Schepler et al. 2022). HUVEC plated on the collagen mats supplemented with Ca-polyP nanoparticles showed a significant increase in cell growth and viability compared to cells in control assays without Ca-polyP particles (Schepler et al. 2022). In the scratch test, these cells showed an increased tendency to migrate on the mats enriched with polyP. Likewise, incorporating the particles into the collagen matrix markedly increased the ability of the HUVEC to penetrate the mats.

4.9 First Human Studies

First clinical studies on patients suffering from chronic wounds have been extremely successful. The results obtained exceeded the expectations. Instead of the Zn-polyP nanoparticles described above, amorphous Ca-polyP nanoparticles have been used, which in turn were integrated into a matrix of compressed collagen (Schepler et al. 2022). A collagen matrix was used, which was produced according to a modified protocol. The formation of collagen subfibrils from the collagen type I molecules dissolved in acetic acid pH 3.6 was again induced by transfer to a citrate-phosphate buffer pH 6, from which the mature collagen fibers were obtained. In order to obtain a more fluffy collagen framework into which the cells can migrate even better, the pressure applied during compression was reduced and the duration of the process shortened. About 3–5 mm thick collagen gels were pressed to about 1 mm thick mats. Furthermore, it turned out to be advantageous to add, during the treatment of the patient, polyP in soluble form to the mats, as Na-polyP which is immediately available as a biologically active polymer, in addition to the Ca-polyP nanoparticles which serve as depot form.

So far, the wound mats have been used on more than 10 patients with nonhealing chronic wounds. All patients recovered completely. As an example, the healing process of a chronic wound on the ventral tibia of the left lower leg of a 79-year-old patient that could not be treated with conventional methods is shown. This wound, approximately 5 cm in diameter, had developed following surgical resection of a squamous cell carcinoma; Fig. 4.8a-A shows the location of the wound as well as the Ca-polyP-nanoparticle-supplemented wound mat of approximately the same size, which completely covered the wound after use (Schepler et al. 2022). After

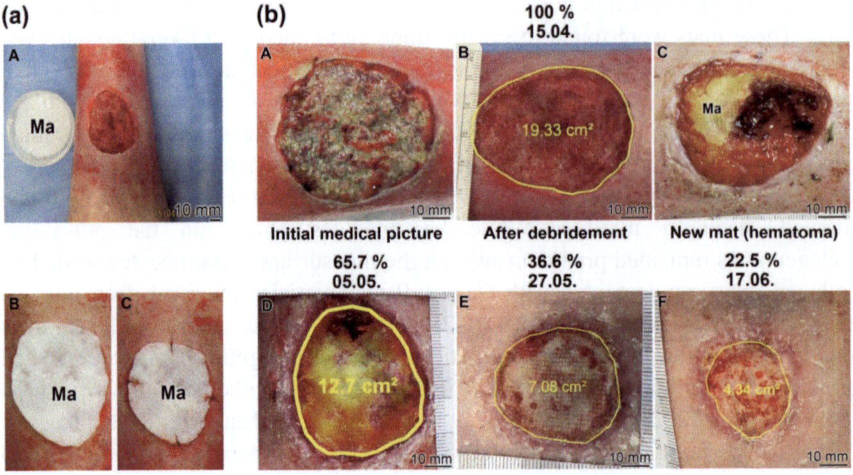

Fig. 4.8 Successful healing of a chronic wound on the ventral tibia of the left lower leg of a 79-year-old patient by applying the Ca-polyP nanoparticle-containing compressed collagen mats. (**a-A**) Petri dish with the mat next to the leg with the wound. (**a-B** and **a-C**) Deposition of the mat on the wound before (**a-B**) and after fixation (**a-C**). (**b-A** to **b-F**) Progress in wound healing. (**b-A**) Wound before debridement. (**b-B**) Wound after debridement. (**b-C**) Change of mat after development of a hematoma. Status of wound healing after 3 weeks (**b-D**), 6 weeks (**b-E**), and 9 weeks (**b-F**). In (**b-B**) and (**b-D**) to (**b-F**) the size of the wound (circled) is given in percent of the initial wound area. Adapted with permission from Schepler et al. (2022). Copyright 2022, Ivyspring International Publisher

debridement (cleaning of the wound by surgery and ultrasound treatment), the mat was applied to the wound and fixed (Fig. 4.8a-B and a-C). The condition of the wound before and after debridement is shown in Fig. 4.8b-A and b-B. The progress of wound healing was then monitored by regularly determining the size of the wound defect. It was found that treating the wound with the Ca-polyP-nanoparticle-supplemented collagen mat impressively accelerated the rate of reepithelialization (Fig. 4.8b-B and Fig. 4.8b-D to b-F). Already after 3 weeks a reduction of the non-epithelialized wound area to 65.7% (12.7 cm^2; Fig. 4.8b-D) was found, and after 6 and 9 weeks to 36.6% (7.08 cm^2; Fig. 4.8b-E) and 22.5% (4.34 cm^2; Fig. 4.8b-F), respectively (Schepler et al. 2022). The wound area before treatment (19.33 cm^2; Fig. 4.8b-B) was set to 100%.

Biopsy samples from the regenerating wound area, taken 7 days after the start of treatment, showed greatly increased cell proliferation and granulation tissue formation, characteristic signs of active wound healing (Schepler et al. 2022). The granulation tissue mainly contained myofibroblasts, but no detectable keratinocytes. Myofibroblasts were found not only on the surface of the wound-facing side of the wound mat, but also up to 250 μm inside the mat, indicating active cell migration (Schepler et al. 2022). In addition to the myofibroblasts, immune cells were detected in the granulation tissue as well as CD31-positive endothelial cells that were arranged around the small blood vessels.

Furthermore, it was found that the polyP nanoparticles on the wound side of the mats quickly transform into a coacervate phase into which the myofibroblasts infiltrate (Schepler et al. 2022). In contrast, no evidence of coacervation was found on the outward facing surface of the mats.

4.10 Alginate-Gelatin-Based Matrix Supplemented with PolyP

In addition, we developed a biomimetic hydrogel matrix for wound treatment that mimics the physiological ECM. This matrix consists of alginate and periodate-oxidized alginate to which gelatin is bound via Schiff base formation. PolyP is then integrated by ionic crosslinking with Zn^{2+} ions (Wang et al. 2020b).

The alginate used as the starting material to make this matrix is a block copolymer of 1,4-linked β-D-mannuronic acid and α-L-guluronic acid. After adding divalent cations to alginate, a relatively stiff hydrogel is obtained, which only allows a limited migration of cells. Therefore, part of the alginate was converted to oxidized alginate. By cleavage of the C2–C3 bond of the sugar residues of the alginate with the help of periodate (Jejurikar et al. 2012), two aldehyde groups per sugar unit are obtained, which can be covalently bound to free amino groups of proteins by Schiff base formation (formation of C=N bonds) (Sarker et al. 2017). We used gelatin which was cross-linked to the periodate oxidized alginate framework via the ε-amino groups of its lysine and hydroxylysine residues (Balakrishnan and Jayakrishnan 2005; Sarker et al. 2014). Furthermore, after periodate oxidation, the resulting aldehyde groups react with the OH groups of neighboring non-oxidized uronic acid units to form cyclic hemiacetals, leading to the formation of covalent interchain linkages, in addition to ionic crosslinking occurring in the presence of divalent cations.

The hydrogel matrix obtained in this way shows improved interaction with cells, but has the disadvantage of not being morphogenetically active and not supplying any metabolic energy. This was achieved by incorporating the Zn-polyP nanoparticles. Two forms of Zn-polyP-containing matrices were produced (Wang et al. 2020b). First, an alginate-oxidized alginate-gelatin matrix that contained the Zn-polyP nanoparticles. The polyP nanoparticles (depot form) contained therein, formed at pH 10, are converted into the biologically active coacervate when they come into contact with protein (wound secretion). Second, a matrix that contained the polyP coacervate, which was directly formed at neutral pH. In both cases, the polyP is incorporated into the matrix both by binding to the carboxylate or OH groups of the alginate via divalent cations (Zn^{2+}) and as a result of the ionic crosslinking of the alginate-gelatine gel.

The experiments revealed that both polyP-containing hydrogel matrices showed a significantly better effect on promotion of the growth, viability, attachment, and energy-dependent migration of keratinocytes compared to cells growing on a polyP-lacking hydrogel matrix (Wang et al. 2020b).

4.11 Application of 3D Printing

The use of 3D bioprinting technology is an attractive possibility for the fabrication of polyP-containing wound healing mats. This technique allows the manufacture of tailor-made wound healing mats that are enriched with polyP nanoparticles. A combination with other manufacturing processes, such as those used to prepare compressed wound mats, is also possible.

The procedure is shown schematically in Fig. 4.9a. In the first step, a cell containing 3D hydrogel mesh is printed using the 3D bioprinting process. In the second step, the 3D bioprinted hydrogel is covered with a compressed collagen mat made using the process described in Sect. 4.7. The wound dressing obtained is thus a two-layer mat. It consists of a stronger and wound-protective outer layer made of compressed collagen and a wound-facing, more voluminous inner layer made of the 3D-printed hydrogel (Fig. 4.9b). Both layers are supplemented with polyP particles. In our approach, human keratinocytes were directly added to the ink used for 3D

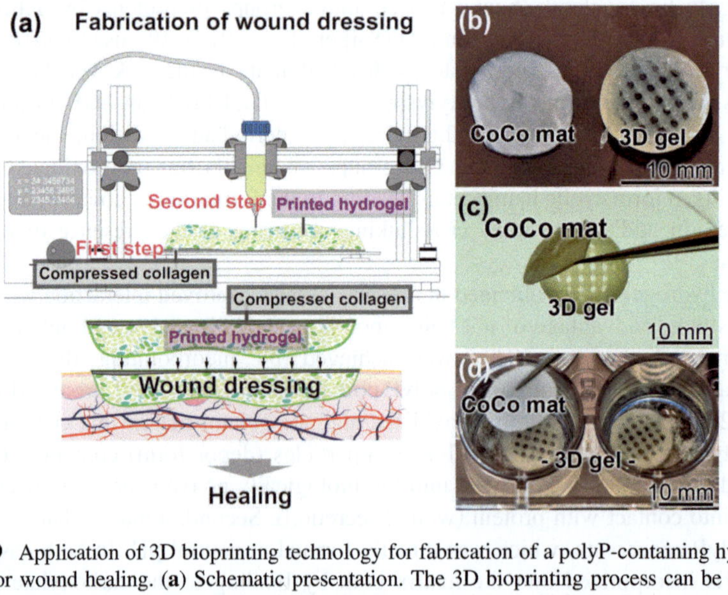

Fig. 4.9 Application of 3D bioprinting technology for fabrication of a polyP-containing hydrogel mesh for wound healing. (**a**) Schematic presentation. The 3D bioprinting process can be used to produce tailor-made hydrogels enriched with polyP nanoparticles, which are then covered with a compressed collagen mat. (**b**) Compressed collagen mat (CoCo mat) and 3D bioprinted hydrogel. (**c**) Applying the mat to the hydrogel. (**d**) Incubation of the 3D bioprinted hydrogel covered with the collagen mat in a cell suspension within a cell culture plate

printing. In these experiments, we used a bio-ink consisting of an *N,O*-carboxymethyl chitosan- (Müller et al. 2015d) and alginate-based hydrogel, which was spiked with Na-polyP and Zn-polyP nanoparticles. After the collagen mat had been applied to the hydrogel (Fig. 4.9c), the resulting two-layer mat with the embedded keratinocytes was placed in a cell culture plate with growth medium and incubated (Fig. 4.9d). The matrix showed a strong stimulating effect on the growth of the keratinocytes embedded in the mats, as well as on keratinocytes seeded on the surface of the mats.

4.12 ALP/ADK: A Bio-artificial Intelligence System?

With the wound mats described here, we have succeeded in introducing a new principle in wound therapy: the use of a polymer that supplies the metabolic energy required for wound healing. The lack of energy in wound healing particularly affects the extracellular space with only low ATP levels. Exogenously administered polyP and the body's own ALP/ADK pair could serve as an example of a biological artificial intelligence (AI) system (Ghahramani 2015; Nesbeth et al. 2016; Zhang et al. 2018), a so-called bio-artificial intelligence (BAI) system, ensuring adequate availability of metabolic energy for tissue repair. There are two systems that are proposed to closely interact with each other, a fast-reacting system based on replenishment of the endogenous ATP pool via polyP (Fig. 4.10a), and a slower-reacting system involving gene induction (induction of ALP) via polyP (Fig. 4.10b).

A disturbance (i.e., ATP deficiency) occurs with tissue damage (skin injury, bone fracture, cartilage degeneration, etc.) that requires metabolic energy in the form of ATP to repair. This energy is supplied by polyP, either endogenous polyP released by platelets migrating to the site of injury, or exogenous polyP, as used here to treat therapy-resistant chronic wounds. The body's own enzyme-based systems then generate ATP from polyP. These systems must be able to sense and respond to the actual energy state, the AMP:ATP ratio, either very quickly (ATP synthesis via the ALP/ADK couple) or in the longer term (via gene/*ALP* induction). Each of these systems forms a circuit, as shown in Fig. 4.10a, b, and the overall system is most likely capable of learning, as required for an AI system.

PolyP plays the central role in both circuits that closely cooperate with each other. This polymer can be considered as a bio-intelligent material. Depending on the stimulus, it releases the metabolic energy that is required for wound healing. This ability of polyP is based on its property of being subject to conversion from a stable nanoparticulate state (amorphous nanoparticles) to a physiologically active and energy-delivering coacervate phase when it comes into contact with protein-containing body fluids such as wound secretion. This triggers the biological activity of the polyP, both its morphogenetic activity and its ability to act as a donor of metabolic energy.

In addition, polyP has the property of tailoring its response to the particular type of tissue that is damaged, with the specific response elicited by polyP, depending on the nature of its counterions. Ca-polyP nanoparticles preferably exhibit morphogenetic activity in bone defects (Müller et al. 2017c), Mg-polyP nanoparticles in cartilage damage (Wang et al. 2018), and Mg-polyP nanoparticles (Müller et al.

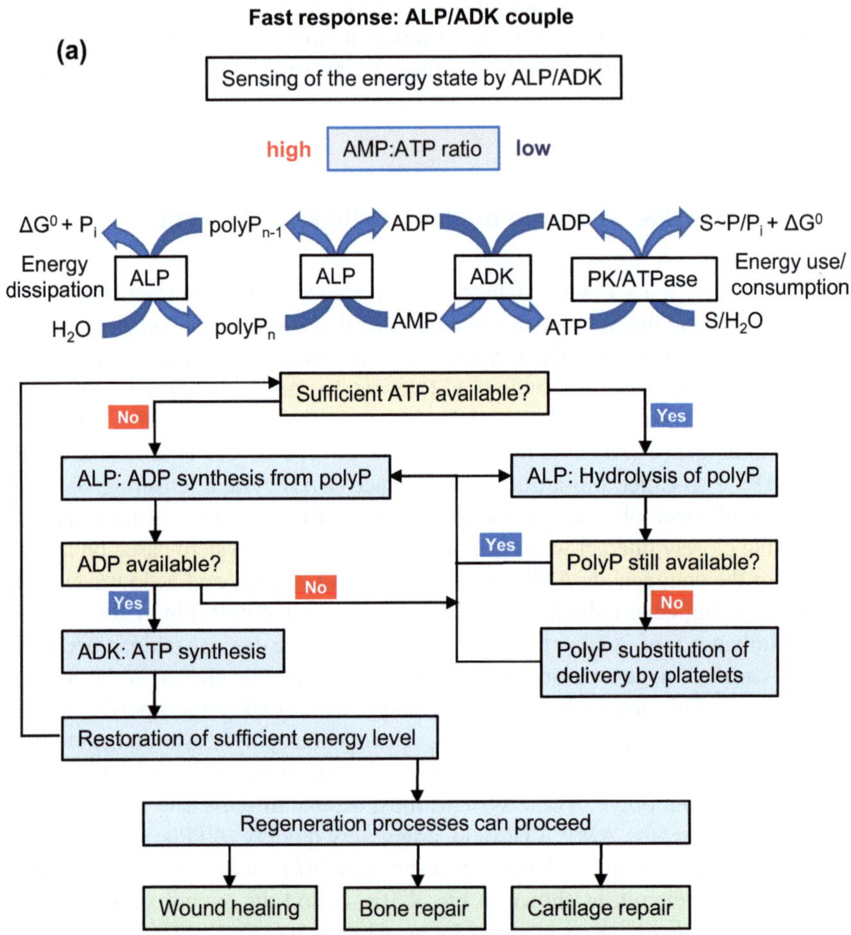

Fig. 4.10 Exogenously administrated polyP and the ALP/ADK couple as an example of a bio-artificial intelligence (BAI) system. The system consists of two parts, which are assumed to closely interact with each other, a fast-responding system, able to replenish the endogenous ATP pool via polyP (**a**), and a slow-responding system, involving induction of the *ALP* gene via polyP (**b**). Perturbations occur in case of tissue damage (skin wounding, bone fracture, cartilage degeneration, etc.), which requires metabolic energy (ATP) for repair. This energy is delivered by polyP, an intelligent biomaterial that can be present either in an "inactive" particular form or an "active" non-particular (soluble) form depending on the external conditions (pH, protein). The body own ALP/ADK system involved in the generation of ATP from polyP is able to sense the actual energy status (AMP:ATP ratio) and to respond both immediately (ATP synthesis via the ALP/ADK couple) and in the long term (via *ALP* gene induction). Each part of the system forms a circuit, and the overall system most likely is capable of learning. Both systems are presented in the form of a yes/no framework

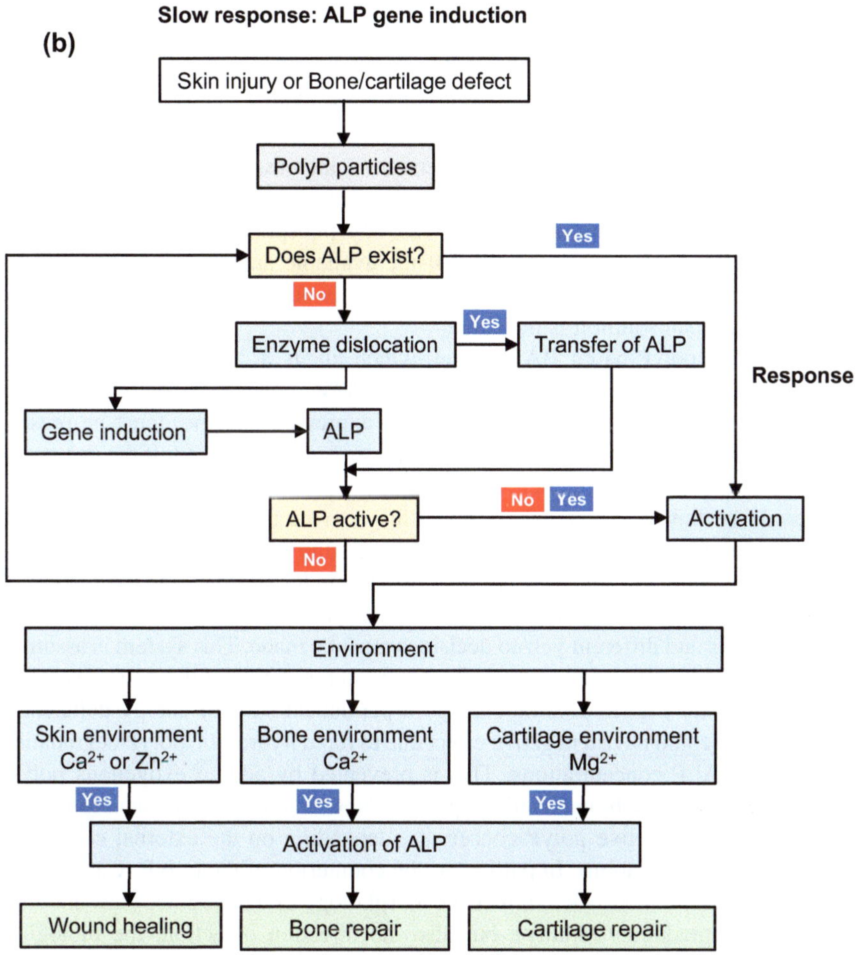

Fig. 4.10 (continued)

2020) or Ca-polyP nanoparticles (Müller et al. 2017b; Schepler et al. 2022) in wounded skin.

In biological systems, AI circuits involve links between a sensor, a decision-making system, and an effector via intelligent biomaterials (Ghahramani 2015; Nesbeth et al. 2016). Such a BAI system can include an intelligent biomaterial such as polyP that is instructive and triggers a specific effect in cells and tissues and whose response is modulated by internal or external stimuli.

In Fig. 4.10a, a scheme of the first part of the polyP-based BAI system, which induces a rapid response to the eliciting stimulus (ATP deficiency), is shown. This system senses the energy state given by the AMP:ATP ratio. The effect is either an increased ATP production, if the AMP:ATP level is high, or the hydrolysis of the polyP, if the AMP:ATP level is low. In the latter case, only energy dissipation occurs. The two key enzymes are the ALP, which becomes active when the AMP levels are elevated, and the ADK, which facilitates the adjustment of the equilibration between the three adenine nucleotides, AMP, ADP, and ATP. Consequently, only the energy charge ([ATP] + ½ [ADP] / [ATP] + [ADP] + [AMP]) increases, but not the total amount of the adenine nucleotides. If enough ATP is present, polyP undergoes hydrolysis. If the AMP concentration increases, the ALP transfers the energy-rich phosphate from polyP to AMP. Only in the case of an increased concentration of ADP (formed via the ALP by phospho-transfer from polyP to AMP) does the ADK regenerate ATP. If the available amount of polyP becomes too small as a result of polyP hydrolysis or if not enough polyP is provided by the platelets, polyP substitution is indicated.

The second polyP-based BAI system, which elicits a delayed gene-induction-based effect triggered by a stimulus (energy/ATP deficiency), is schematically shown in Fig. 4.10b. The ALP, as an active, expressed enzyme, must be present for the effect to occur. Only then metabolic energy can be produced. If the enzyme is present but only little active, then activation of the enzyme by polyP can take place. This could be linked to a translocation of the enzyme from the membrane-bound state to a soluble state or vice versa (Rader 2017). If the ALP is missing or if there are insufficient amounts of the enzyme, then enzyme induction, i.e., the expression of the *ALP* gene, must be triggered (Wang et al. 2016d). Thus, regulation is possible at different levels and different yes/no decisions must be made. This system is assumed also to be capable of learning, i.e., a long-lasting upregulation of the ALP is possible with a frequently triggered stimulus, e.g., a permanent state of energy deficiency. However, this could have a negative effect due to rapid hydrolytic polyP degradation at very high ALP concentrations. This is prevented by adding exogenous polyP. PolyP, as an intelligent biomolecule, is present either as stable polyP nanoparticles or as an immediately active polyP coacervate, depending on the external conditions (pH, protein concentration). In particular, the counterion of the polyP (Ca^{2+}, Mg^{2+}, Zn^{2+}, and others) not only determines the morphology of the particles (size and thus the rate of degradation/stability), but also the direction of action, the biological effect, as described above (Müller et al. 2016; Wang et al. 2016a, b, d; Hu et al. 2018). In the case of disturbances of wound healing, a Zn^{2+} deficiency could be a decisive factor. Zn^{2+} accumulates during wound healing (Kogan et al. 2017). In addition to Mg^{2+}, Zn^{2+} is a cofactor of the ALP and essential for the activity of the enzyme (Sorimachi 1987).

Future studies need to substantiate the evidence that polyP is part of a BAI system and how the different components of this system work together, in particular, whether and how this system is equipped with a memory and a learning ability.

Acknowledgments W.E.G.M. is a holder of an ERC Advanced Investigator Grant (No. 268476). In addition, W.E.G.M. obtained three ERC-PoC grants (Si-Bone-PoC, No. 324564; MorphoVES-PoC, No. 662486; and ArthroDUR, No. 767234). We also acknowledge funding from the European Commission (grants BIO-SCAFFOLDS No. 604036 and BlueGenics No. 311848). Finally, this work was supported by a grant from the Federal Minister of Education and Research (No. 13GW0403B) and the BiomaTiCS research initiative of the University Medical Center, Mainz.

References

Abdel Malik R, Zippel N, Frömel T, Heidler J, Zukunft S, Walzog B, Ansari N, Pampaloni F, Wingert S, Rieger MA, Wittig I, Fisslthaler B, Fleming I (2017) AMP-activated protein kinase α2 in neutrophils regulates vascular repair *via* hypoxia-inducible factor-1α and a network of proteins affecting metabolism and apoptosis. Circ Res 120:99–109

Agren MS (1993) Zinc oxide increases degradation of collagen in necrotic wound tissue. Br J Dermatol 129:221–222

Atkinson DE (1968) The energy charge of the adenylate pool as a regulatory parameter. Interaction with feedback modifiers. Biochemistry 7:4030–4034

Bae WJ, Auh QS, Kim GT, Moon JH, Kim EC (2016) Effects of sodium tri- and hexametaphosphate *in vitro* osteoblastic differentiation in periodontal ligament and osteoblasts, and *in vivo* bone regeneration. Differentiation 92:257–269

Balakrishnan B, Jayakrishnan A (2005) Self-cross-linking biopolymers as injectable *in situ* forming biodegradable scaffolds. Biomaterials 26:3941–3951

Barysch MJ, Läuchli S (2020) Oxygen therapy in wound healing. In: Alavi A, Maibach H (eds) Local wound care for dermatologists. Updates in clinical dermatology. Springer, Cham. https://doi.org/10.1007/978-3-030-28872-3_12

Beidler SK, Douillet CD, Berndt DF, Keagy BA, Rich PB, Marston WA (2009) Inflammatory cytokine levels in chronic venous insufficiency ulcer tissue before and after compression therapy. J Vasc Surg 49:1013–1020

Carling D, Zammit VA, Hardie DG (1987) A common bicyclic protein kinase cascade inactivates the regulatory enzymes of fatty acid and cholesterol biosynthesis. FEBS Lett 223:217–222

Chattopadhyay S, Raines RT (2014) Collagen-based biomaterials for wound healing. Biopolymers 101:821–833

Chicharro-Alcántara D, Rubio-Zaragoza M, Damiá-Giménez E, Carrillo-Poveda JM, Cuervo-Serrato B, Peláez-Gorrea P, Sopena-Juncosa JJ (2018) Platelet rich plasma: new insights for cutaneous wound healing management. J Funct Biomater 9:10

Docampo R, Ulrich P, Moreno SN (2010) Evolution of acidocalcisomes and their role in polyphosphate storage and osmoregulation in eukaryotic microbes. Philos Trans R Soc Lond Ser B Biol Sci 365:775–784

Ghahramani Z (2015) Probabilistic machine learning and artificial intelligence. Nature 521:452–459

Gosain A, DiPietro LA (2004) Aging and wound healing. World J Surg 28:321–326

Guo S, DiPietro LA (2010) Factors affecting wound healing. J Dent Res 89:219–229

Han G, Ceilley R (2017) Chronic wound healing: a review of current management and treatments. Adv Ther 34:599–610

Hardie DG (2004) AMP-activated protein kinase: the guardian of cardiac energy status. J Clin Invest 114:465–468

Hardie DG (2007) AMP-activated/SNF1 protein kinases: conserved guardians of cellular energy. Nat Rev Mol Cell Biol 8:774–785

Hardie DG, Hawley SA, Scott JW (2006) AMP-activated protein kinase: development of the energy sensor concept. J Physiol 574:7–15

Hardie DG, Schaffer BE, Brunet A (2016) AMPK: an energy-sensing pathway with multiple inputs and outputs. Trends Cell Biol 26:190–201

Hu T, Xu H, Wang C, Qin H, An Z (2018) Magnesium enhances the chondrogenic differentiation of mesenchymal stem cells by inhibiting activated macrophage-induced inflammation. Sci Rep 8:3406

Huang H, Wang L, Qian F, Chen X, Zhu H, Yang M, Zhang C, Chu M, Wang X, Huang X (2021) Liraglutide *via* activation of AMP-activated protein kinase-hypoxia inducible factor-1α-heme oxygenase-1 signaling promotes wound healing by preventing endothelial dysfunction in diabetic mice. Front Physiol 12:660263

Im MJ, Hoopes JE (1970) Energy metabolism in healing skin wounds. J Surg Res 10:459–464

Jang WG, Kim EJ, Lee KN, Son HJ, Koh JT (2011) AMP-activated protein kinase (AMPK) positively regulates osteoblast differentiation via induction of Dlx5-dependent Runx2 expression in MC3T3E1 cells. Biochem Biophys Res Commun 404:1004–1009

Jejurikar A, Seow XT, Lawrie G, Martin D, Jayakrishnan A, Grøndahl (2012) Degradable alginate hydrogels crosslinked by the macromolecular crosslinker alginate dialdehyde. J Mater Chem 22: 9751–9758

Karu TI (2010) Multiple roles of cytochrome c oxidase in mammalian cells under action of red and IR-A radiation. IUBMB Life 62:607–610

Kaushika M, Niranjana R, Ramar T, Balaraman M, Pandiyarasan V, Ramachandran C, Oh DH, Devan GV (2019) Investigations on the antimicrobial activity and wound healing potential of ZnO nanoparticles. Appl Surf Sci 479:1169–1177

Kiritsi D, Nyström A (2018) The role of TGFβ in wound healing pathologies. Mech Ageing Dev 172:51–58

Kobayashi K, Maeda K, Takefuji M, Kikuchi R, Morishita Y, Hirashima M, Murohara T (2017) Dynamics of angiogenesis in ischemic areas of the infarcted heart. Sci Rep 7:7156

Kogan S, Sood A, Garnick MS (2017) Zinc and wound healing: a review of zinc physiology and clinical applications. Wounds 29:102–106

Kotwal GJ, Sarojini H, Chien S (2015) Pivotal role of ATP in macrophages fast tracking wound repair and regeneration. Wound Repair Regen 23:724–727

Lansdown AB (1993) Influence of zinc oxide in the closure of open skin wounds. Int J Cosmet Sci 15:83–85

Lansdown AB, Mirastschijski U, Stubbs N, Scanlon E, Agren MS (2007) Zinc in wound healing: theoretical, experimental, and clinical aspects. Wound Repair Regen 15:2–16

Leyane TS, Jere SW, Houreld NN (2021) Cellular signalling and photobiomodulation in chronic wound repair. Int J Mol Sci 22:11223

Lin JT, Chen HM, Chiu CH, Liang YJ (2014) AMP-activated protein kinase activators in diabetic ulcers: from animal studies to phase II drugs under investigation. Expert Opin Investig Drugs 23:1253–1265

Lin PH, Sermersheim M, Li H, Lee PHU, Steinberg SM, Ma J (2017) Zinc in wound healing modulation. Nutrients 10:pii: E16

Lorenz B, Schröder HC (2001) Mammalian intestinal alkaline phosphatase acts as highly active exopolyphosphatase. Biochim Biophys Acta 1547:254–261

Martin YH, Jubin K, Smalley S, Wong JPF, Brown RA, Metcalfe AD (2017) A novel system for expansion and delivery of human keratinocytes for the treatment of severe cutaneous injuries using microcarriers and compressed collagen. J Tissue Eng Regen Med 11:3124–3133

Mirastschijski U, Haaksma CJ, Tomasek JJ, Agren MS (2004) Matrix metalloproteinase inhibitor GM 6001 attenuates keratinocyte migration, contraction and myofibroblast formation in skin wounds. Exp Cell Res 299:465–475

Mo Y, Sarojini H, Wan R, Zhang Q, Wang J, Eichenberger S, Kotwal GJ, Chien S (2020) Intracellular ATP delivery causes rapid tissue regeneration *via* upregulation of cytokines, chemokines, and stem cells. Front Pharmacol 10:1502

Moehring F, Cowie AM, Menzel AD, Weyer AD, Grzybowski M, Arzua T, Geurts AM, Palygin O, Stucky CL (2018) Keratinocytes mediate innocuous and noxious touch via ATP-P2X4 signaling. elife 7:pii:e31684

Morrissey JH, Choi SH, Smith SA (2012) Polyphosphate: an ancient molecule that links platelets, coagulation, and inflammation. Blood 119:5972–5979

Mosser DM, Edwards JP (2008) Exploring the full spectrum of macrophage activation. Nat Rev Immunol 8:958–969

Müller WEG, Wang XH, Diehl-Seifert B, Kropf K, Schloßmacher U, Lieberwirth I, Glasser G, Wiens M, Schröder HC (2011) Inorganic polymeric phosphate/polyphosphate is an inducer of alkaline phosphatase and a modulator of intracellular Ca^{2+} level in osteoblasts (SaOS-2 cells) *in vitro*. Acta Biomater 7:2661–2671

Müller WEG, Tolba E, Dorweiler B, Schröder HC, Diehl-Seifert B, Wang XH (2015a) Electrospun bioactive mats enriched with Ca-polyphosphate/retinol nanospheres as potential wound dressing. Biochem Biophys Rep 3:150–160

Müller WEG, Tolba E, Feng Q, Schröder HC, Markl JS, Kokkinopoulou M, Wang XH (2015b) Amorphous Ca^{2+} polyphosphate nanoparticles regulate the ATP level in bone-like SaOS 2 cells. J Cell Sci 128:2202–2207

Müller WEG, Tolba E, Schröder HC, Diehl-Seifert B, Wang XH (2015c) Retinol encapsulated into amorphous Ca^{2+} polyphosphate nanospheres acts synergistically in MC3T3-E1 cells. Eur J Pharm Biopharm 93:214–223

Müller WEG, Tolba E, Schröder HC, Neufurth M, Wang SF, Link T, Al-Nawas B, Wang XH (2015d) A new printable and durable *N,O*-carboxymethyl chitosan-Ca^{2+}-polyphosphate complex with morphogenetic activity. J Mater Chem B 3:1722–1730

Müller WEG, Tolba E, Schröder HC, Wang XH (2015e) Polyphosphate: a morphogenetically active implant material serving as metabolic fuel for bone regeneration. Macromolec Biosci 15: 1182–1197

Müller WEG, Tolba E, Schröder HC, Wang SF, Glaßer G, Muñoz-Espí R, Link T, Wang XH (2015f) A new polyphosphate calcium material with morphogenetic activity. Mater Lett 148: 163–166

Müller WEG, Ackermann M, Tolba E, Neufurth M, Wang S, Schröder HC, Wang XH (2016) A bio-imitating approach to fabricate an artificial matrix for cartilage tissue engineering using magnesium-polyphosphate and hyaluronic acid. RSC Adv 6:88559–88570

Müller WEG, Neufurth M, Tolba E, Ackermann M, Korzhev M, Wang S, Feng Q, Schröder HC, Wang XH (2017a) Bifunctional dentifrice: amorphous polyphosphate a regeneratively active sealant with potent anti-*Streptococcus mutans* activity. Dent Mater 33:753–764

Müller WEG, Relkovic D, Ackermann M, Wang S, Neufurth M, Paravic-Radicevic A, Ushijima H, Schröder HC, Wang XH (2017b) Enhancement of wound healing in normal and diabetic mice by topical application of amorphous polyphosphate – superior effect of the host-guest composite material composed of collagen (host) and polyphosphate (guest). Polymers 9:300

Müller WEG, Tolba E, Ackermann M, Neufurth M, Wang S, Feng Q, Schröder HC, Wang XH (2017c) Fabrication of amorphous strontium polyphosphate microparticles that induce mineralization of bone cells *in vitro* and *in vivo*. Acta Biomater 50:89–101

Müller WEG, Wang SF, Ackermann M, Neufurth M, Steffen R, Mecja E, Muñoz-Espí R, Feng QL, Schröder HC, Wang XH (2017d) Rebalancing β-amyloid-induced decrease of ATP level by amorphous nano/micro polyphosphate: suppression of the neurotoxic effect of amyloid β-protein fragment 25-35. Int J Mol Sci 18:2154

Müller WEG, Wang SF, Neufurth M, Kokkinopoulou M, Feng Q, Schröder HC, Wang XH (2017e) Polyphosphate as a donor of high-energy phosphate for the synthesis of ADP and ATP. J Cell Sci 130:2747–2756

Müller WEG, Wang SF, Wiens M, Neufurth M, Ackermann M, Relkovic D, Kokkinopoulou M, Feng Q, Schröder HC, Wang XH (2017f) Uptake of polyphosphate microparticles *in vitro* (SaOS-2 and HUVEC cells) followed by an increase of the intracellular ATP pool size. PLoS One 12(12):e0188977

Müller WEG, Ackermann M, Tolba E, Neufurth M, Ivetac I, Kokkinopoulou M, Schröder HC, Wang XH (2018a) Role of ATP during the initiation of microvascularization. Acceleration of an autocrine sensing mechanism facilitating chemotaxis by inorganic polyphosphate. Biochemist J 3255–3273

Müller WEG, Ackermann M, Wang SF, Neufurth M, Muñoz-Espí R, Feng QL, Schröder HC, Wang XH (2018b) Inorganic polyphosphate induces accelerated tube formation of HUVEC endothelial cells. Cell Mol Life Sci 75:21–32

Müller WEG, Wang S, Tolba E, Neufurth M, Ackermann M, Muñoz-Espí R, Lieberwirth I, Glasser G, Schröder HC, Wang XH (2018c) Transformation of amorphous polyphosphate nanoparticles into coacervate complexes: an approach for the encapsulation of mesenchymal stem cells. Small 14:e1801170

Müller WEG, Schröder HC, Wang XH (2019a) Inorganic polyphosphates as storage for and generator of metabolic energy in the extracellular matrix. Chem Rev 119:12337–12374
Müller WEG, Schröder HC, Wang XH (2019b) The phosphoanhydride bond: one cornerstone of life. Biochem (Lond) 41:22–27
Müller WEG, Schepler H, Tolba E, Wang SF, Ackermann M, Muñoz-Espí R, Xiao S, Tan RW, She ZD, Neufurth M, Schröder HC, Wang XH (2020) A physiologically active interpenetrating collagen network that supports growth and migration of epidermal keratinocytes: zinc-polyP nanoparticles integrated into compressed collagen. J Mater Chem B 8:5892–5902
Nesbeth DN, Zaikin A, Saka Y, Romano MC, Giuraniuc CV, Kanakov O, Laptyeva T (2016) Synthetic biology routes to bio-artificial intelligence. Essays Biochem 60:381–391
Neufurth M, Wang XH, Wang SF, Steffen R, Ackermann M, Haep ND, Schröder HC, Müller WEG (2017) 3D printing of hybrid (bio)materials for bone tissue engineering: calcium-polyphosphate microparticles encapsulated by polycaprolactone. Acta Biomater 64:377–388
Nurden AT (2018) The biology of the platelet with special reference to inflammation, wound healing and immunity. Front Biosci (Landmark Ed) 23:726–751
Oakhill JS, Scott JW, Kemp BE (2009) Structure and function of AMP-activated protein kinase. Acta Physiol (Oxf) 196:3–14
Oakhill JS, Steel R, Chen ZP, Scott JW, Ling N, Tam S, Kemp BE (2011) AMPK is a direct adenylate charge-regulated protein kinase. Science 332:1433–1435
Omura J, Satoh K, Kikuchi N, Satoh T, Kurosawa R, Nogi M, Otsuki T, Kozu K, Numano K, Suzuki K, Sunamura S, Tatebe S, Aoki T, Sugimura K, Miyata S, Hoshikawa Y, Okada Y, Shimokawa H (2016) Protective roles of endothelial AMP-activated protein kinase against hypoxia-induced pulmonary hypertension in mice. Circ Res 119:197–209
Paneni F, Beckman JA, Creager MA, Cosentino F (2013) Diabetes and vascular disease: pathophysiology, clinical consequences, and medical therapy: part I. Eur Heart J 34:2436–2443
Pavlica S, Gaunitz F, Gebhardt R (2009) Comparative *in vitro* toxicity of seven zinc-salts towards neuronal PC12 cells. Toxicol In Vitro 23:653–659
Rader BA (2017) Alkaline phosphatase, an unconventional immune protein. Front Immunol 8:897
Reinke JM, Sorg H (2012) Wound repair and regeneration. Eur Surg Res 49:35–43
Rittié L (2016) Cellular mechanisms of skin repair in humans and other mammals. J Cell Commun Signal 10:103–120
Rodríguez C, Muñoz M, Contreras C, Prieto D (2021) AMPK, metabolism, and vascular function. FEBS J 288:3746–3771
Rolfe DF, Brown GC (1997) Cellular energy utilization and molecular origin of standard metabolic rate in mammals. Physiol Rev 77:731–758
Saghazadeh S, Rinoldi C, Schot M, Kashaf SS, Sharifi F, Jalilian E, Nuutila K, Giatsidis G, Mostafalu P, Derakhshandeh H, Yue K, Swieszkowski W, Memic A, Tamayol A, Khademhosseini A (2018) Drug delivery systems and materials for wound healing applications. Adv Drug Deliv Rev 127:138–166
Sarker B, Papageorgiou DG, Silva R, Zehnder T, Gul-E-Noor F, Bertmer M, Kaschta J, Chrissafis K, Detsch R, Boccaccini AR (2014) Fabrication of alginate-gelatin crosslinked hydrogel microcapsules and evaluation of the microstructure and physico-chemical properties. J Mater Chem B 2:1470–1482
Sarker B, Singh R, Zehnder T, Forgber T, Alexiou C, Cicha I, Detsch R, Boccaccini AR (2017) Macromolecular interactions in alginate-gelatin hydrogels regulate the behavior of human fibroblasts. J Bioact Compat Polym 32:309–324
Sarojini H, Billeter AT, Eichenberger S, Druen D, Barnett R, Gardner SA, Galbraith NJ, Polk HC Jr, Chien S (2017) Rapid tissue regeneration induced by intracellular ATP delivery-a preliminary mechanistic study. PLoS One 12:e0174899
Schepler H, Neufurth M, Wang SF, She Z, Schröder HC, Wang XH, Müller WEG (2022) Acceleration of chronic wound healing by bio-inorganic polyphosphate: *in vitro* studies and first clinical applications. Theranostics 12:18–34

Schröder HC, Tolba E, Diehl-Seifert B, Wang XH, Müller WEG (2017) Electrospinning of bioactive wound healing nets. In: Müller WEG, Schröder HC, Wang XH (eds) BlueGenics – from gene to bioactive product: exploiting marine genomics for an innovative and sustainable European blue biotechnology industry. Springer, Berlin. Prog Mol Subcell Biol 55:259–290
Schröder HC, Wang XH, Müller WEG (2019) Amorphous polyphosphate nanoparticles: application of the morphogenetically active inorganic polymer for personalized tissue regeneration. J Phys D Appl Phys 52:363001
Simbulan-Rosenthal CM, Gaur A, Sanabria VA, Dussan LJ, Saxena R, Schmidt J, Kitani T, Chen YS, Rahim S, Uren A, Crooke E, Rosenthal DS (2015) Inorganic polyphosphates are important for cell survival and motility of human skin keratinocytes. Exp Dermatol 24:636–639
Singer AJ, Clark RA (1999) Cutaneous wound healing. N Engl J Med 341:738–746
Sorimachi K (1987) Activation of alkaline phosphatase with Mg^{2+} and Zn^{2+} in rat hepatoma cells. Accumulation of apoenzyme. J Biol Chem 262:1535–1541
Takeo M, Lee W, Ito M (2015) Wound healing and skin regeneration. Cold Spring Harb Perspect Med 5:a023267
Tandara AA, Mustoe TA (2004) Oxygen in wound healing—more than a nutrient. World J Surg 28: 294–300
Tousoulis D, Papageorgiou N, Androulakis E, Siasos G, Latsios G, Tentolouris K, Stefanadis C (2013) Diabetes mellitus-associated vascular impairment: novel circulating biomarkers and therapeutic approaches. J Am Coll Cardiol 62:667–676
Trautmann A (2009) Extracellular ATP in the immune system: more than just a “danger signal”. Sci Signal 2(pe6)
Trefts E, Shaw RJ (2021) AMPK: restoring metabolic homeostasis over space and time. Mol Cell 81:3677–3690
Ungvari Z, Tarantini S, Donato AJ, Galvan V, Csiszar A (2018) Mechanisms of vascular aging. Circ Res 123:849–867
Wang XH, Ackermann M, Tolba E, Neufurth M, Wurm F, Feng Q, Wang S, Schröder HC, Müller WEG (2016a) Artificial cartilage bio-matrix formed of hyaluronic acid and Mg^{2+}-polyphosphate. Eur Cell Mater 32:271–283
Wang XH, Ackermann M, Wang S, Tolba E, Neufurth M, Feng Q, Schröder HC, Müller WEG (2016b) Amorphous polyphosphate/amorphous calcium carbonate implant material with enhanced bone healing efficacy in a critical-size-defect in rats. Biomed Mater 11:035005
Wang XH, Schröder HC, Müller WEG (2016c) Polyphosphate as a metabolic fuel in Metazoa: a foundational breakthrough invention for biomedical applications. Biotechnol J 11:11–30
Wang XH, Wang S, He F, Tolba E, Schröder HC, Diehl-Seifert B, Müller WEG (2016d) Polyphosphate as a bioactive and biodegradable implant material: induction of bone regeneration in rats. Adv Engin Mat 18:1406–1417
Wang XH, Ackermann M, Neufurth M, Wang SF, Li Q, Feng QL, Schröder HC, Müller WEG (2017) Restoration of impaired metabolic energy balance (ATP pool) and tube formation potential of endothelial cells under “high glucose”, diabetic conditions by the bioinorganic polymer polyphosphate. Polymers 9:575
Wang XH, Schröder HC, Müller WEG (2018) Amorphous polyphosphate, a smart bioinspired nano–/bio-material for bone and cartilage regeneration: towards a new paradigm in tissue engineering. J Mat Chem B 6:2385–2412
Wang J, Li J, Song D, Ni J, Ding M, Huang J, Yan M (2020a) AMPK: implications in osteoarthritis and therapeutic targets. Am J Transl Res 12:7670–7681
Wang SF, Wang XH, Neufurth M, Tolba E, Schepler H, Xiao S, Schröder HC, Müller WEG (2020b) Biomimetic alginate/gelatin cross-linked hydrogels supplemented with polyphosphate for wound healing applications. Molecules 25:5210
Wang Z, Shen J, Feng E, Jiao Y (2021) AMPK as a potential therapeutic target for intervertebral disc degeneration. Front Mol Biosci 8:789087
Yan Y, Zhou XE, Xu HE, Melcher K (2018) Structure and physiological regulation of AMPK. Int J Mol Sci 19:3534

Yegutkin GG (2008) Nucleotide- and nucleoside-converting ectoenzymes: important modulators of purinergic signalling cascade. Biochim Biophys Acta 1783:673–694

Yi D, Yu H, Lu K, Ruan C, Ding C, Tong L, Zhao X, Chen D (2021) AMPK signaling in energy control, cartilage biology, and osteoarthritis. Front Cell Dev Biol 9:696602

Yin J, Xu K, Zhang J, Kumar A, Yu FS (2007) Wound-induced ATP release and EGF receptor activation in epithelial cells. J Cell Sci 120:815–825

Yokouchi M, Atsugi T, Logtestijn MV, Tanaka RJ, Kajimura M, Suematsu M, Furuse M, Amagai M, Kubo A (2016) Epidermal cell turnover across tight junctions based on Kelvin's tetrakaidecahedron cell shape. elife 5:e19593

Yu S, Sathi GA, Yamamoto O (2016) *In vivo* evaluation of wound healing property of zinc smectite using a rat model. J Ceramic Soc Jpn 124:1199–1204

Zhang K, Wang S, Zhou C, Cheng L, Gao X, Xie X, Sun J, Wang H, Weir MD, Reynolds MA, Zhang N, Bai Y, Xu HHK (2018) Advanced smart biomaterials and constructs for hard tissue engineering and regeneration. Bone Res 6:31

Chapter 5
Biomimetic Polyphosphate Materials: Toward Application in Regenerative Medicine

Heinz C. Schröder, Xiaohong Wang, Meik Neufurth, Shunfeng Wang, and Werner E. G. Müller

Abstract In recent years, inorganic polyphosphate (polyP) has attracted increasing attention as a biomedical polymer or biomaterial with a great potential for application in regenerative medicine, in particular in the fields of tissue engineering and repair. The interest in polyP is based on two properties of this physiological polymer that make polyP stand out from other polymers: polyP has morphogenetic activity by inducing cell differentiation through specific gene expression, and it functions as an energy store and donor of metabolic energy, especially in the extracellular matrix or in the extracellular space. No other biopolymer applicable in tissue regeneration/repair is known that is endowed with this combination of properties. In addition, polyP can be fabricated both in the form of a biologically active coacervate and as biomimetic amorphous polyP nano/microparticles, which are stable and are activated by transformation into the coacervate phase after contact with protein/body fluids. PolyP can be used in the form of various metal salts and in combination with various hydrogel-forming polymers, whereby (even printable) hybrid materials with defined porosities and mechanical and biological properties can be produced, which can even be loaded with cells for 3D cell printing or with drugs and support the growth and differentiation of (stem) cells as well as cell migration/microvascularization. Potential applications in therapy of bone, cartilage and eye disorders/injuries and wound healing are summarized and possible mechanisms are discussed.

H. C. Schröder · X. Wang · M. Neufurth · S. Wang · W. E. G. Müller (✉)
ERC Advanced Investigator Group, Institute for Physiological Chemistry, University Medical Center of the Johannes Gutenberg University, Mainz, Germany
e-mail: hschroed@uni-mainz.de; wang013@uni-mainz.de; wmueller@uni-mainz.de

W. E. G. Müller et al. (eds.), *Inorganic Polyphosphates*, Progress in Molecular and Subcellular Biology 61, https://doi.org/10.1007/978-3-031-01237-2_5

Keywords Biomaterial · Hydrogel · Polyphosphate · Coacervate · Nanoparticle · Energy storage · Morphogenetic activity · Polyanion · Extracellular matrix · Tissue regeneration

5.1 Introduction

Phosphorus, the second element of group 15 in the periodic table, has an essential function in life. This is due to the fact that the oxyanion of phosphorus, the phosphate anion, is able to form high-energy phosphoanhydride bonds. Although these phosphoanhydride linkages are thermodynamically unstable in aqueous solution and the chemical equilibrium is almost exclusively on the side of the monomeric hydrolysis product, these bonds are kinetically stable, a prerequisite for their function in the storage and transfer of metabolically useful energy, mostly leading to the formation of phosphate ester bonds, a process that requires the involvement of enzymes to reduce the activation energy (Walsh 2020). It is assumed that the kinetic stability of these bonds is due to the negative charges of the oxygens of the linked phosphate units at physiological pH, which protect the phosphoanhydride from nucleophilic attack by water molecules (Walsh 2020). Because of the indispensable function of phosphate in energy metabolism, phosphate-containing compounds with energy-rich phosphoanhydride bonds, in particular the polymeric inorganic polyphosphate (polyP), have been assigned an important role in the origin of life (Lazcano and Miller 1996), in addition to their ability/function in coacervate (protocell) formation/compartmentation (Müller et al. 2019b). Nonetheless, phosphorus is an element that is not abundant on Earth and is likely derived from phosphorus-rich stars (Cescutti et al. 2012; Masseron et al. 2020). Indeed, phosphorus is a deficient element, and given the widespread use of phosphate-based compounds in agriculture and households, the development of strategies to reuse phosphorus from wastewater, e.g., through enhanced biological phosphorus removal (Roy et al. 2021), has been recognized as a major public issue.

The turnover of biomolecules with phosphoanhydride bonds is immense. This applies in particular to the universal energy carrier molecule adenosine triphosphate (ATP). The amount of ATP synthesized by an adult human, at 70–80 kg per day, roughly corresponds to the body weight (Milo and Phillips 2015). A high turnover rate has also been reported for polyP, which is subject to a fast anabolism and catabolism, depending on the metabolic situation (Kumble and Kornberg 1995).

Despite its fundamental role in cellular energy metabolism, the greatest amount of phosphate in humans is found extracellularly, fixed in the mineralized bone, about one percent of the body mass, mainly as calcium phosphate in the form of hydroxyapatite (HA) crystals (Hughes et al. 2019; Walsh 2020). Also, bone is constantly being built up and broken down (remodeling). Recent studies have shown that polyP plays a crucial role in mineralization (for a review, see Müller et al. 2018c; Wang et al. 2018).

In addition to phosphate, there are two further oxyanions, which are involved in the formation of biomineral skeletons: silicate and carbonate. Like phosphate, either as amorphous calcium phosphate or as crystalline HA, silica (silicon dioxide formed from silicate; only amorphous) and calcium carbonate (both amorphous and crystalline [aragonite, vaterite, and calcite]) are the building blocks of the skeletons in the animal kingdom. However, only phosphate, but not silicate, forms kinetically stable anhydride bonds that can be used for metabolic energy (ATP) production. Moreover, the pK_a values of orthophosphate (pK_a 2.2, 7.2, and 12.4; Hunter 2012) and orthosilicate (lowest pK_a, 9.5; Hunter 2012) are quite different. In contrast to polyP, which is stable in neutral aqueous solutions, polysilicic acid is only stable at alkaline pH (Dietzel 2000). Also, the anhydride bond formed between arsenate, the oxyanion of arsenic, the next group 15 element after phosphorus, is less stable than the one between phosphate, as are the arsenate diesters, which, in contrast to the phosphodiester bonds in RNA and DNA, are very unstable (Fekry et al. 2011). The only natural arsenic compounds are known from sponges such as the sponge metabolite arsenicin A, which contains arsenic acid in the [+3] oxidation state, in contrast to phosphorus which is almost exclusively in the [+5] oxidation state (Mancini et al. 2006; Mancini and Defant 2013).

Because of their prominent role in mineralization/skeletal formation and energy metabolism, phosphate and its polymers are promising materials for regenerative medicine. In this chapter, the physiological function of polyP, both extracellularly and intracellularly, and its unrivalled morphogenetic and metabolic energy providing properties are described, which make this polymer a unique "biomedical inorganic polymer" (Müller et al. 2013b) in tissue engineering and repair.

5.2 Extracellular Matrix and Principles of Biomineralization as a Blueprint for Biomimetic Implant Materials

5.2.1 Extracellular Matrix Structure

With regard to the design of a successful tissue engineering or tissue implant material, it seems reasonable to imitate the building principles of the physiological extracellular matrix (ECM), in which the cells are embedded. The ECM consists of a hydrogel scaffold, containing a dynamic, hierarchically organized structural network of functionally active biopolymers (Frantz et al. 2010). Both under normal and pathological conditions, the ECM is subject to permanent remodeling, especially during tissue repair (Theocharis et al. 2019). In many human tissues, the ECM is very voluminous and the space occupied by the cells is very small. For example, calcified bone only consists of 2–5% cells, besides 70% bone mineral, 25% organic matrix (proteoglycans, glycosaminoglycans, and collagen), and 5% water (Upadhyay 2017). Even lower is the cell volume in articular cartilage with about 1–2% of cells (Quinn et al. 2013), which belongs to the bradytrophic tissues that lack any vascularization and solely depend on the diffusion of nutrients from the surrounding environment. In general, the concentration of ATP in the extracellular space is low, but the formation and maintenance of the function and structural organization of the ECM is a strongly metabolic energy consuming process. This energy is needed not only for the synthesis of the structural macromolecules such as collagen or elastin that build the ECM scaffold, as well as the hydrogel-forming glycosaminoglycans and proteoglycans such as chondroitin sulfate and hyaluronic acid in bone or aggrecan in cartilage, but also for the assembly of these biopolymers, as well as for the activity of extracellular chaperones (Poon et al. 2000; Müller et al. 2017f) and kinases (Tagliabracci et al. 2012, 2013) to maintain the ECM function (for a discussion, see Müller et al. 2019a).

5.2.2 *Role of Enzymes in Biomineral Formation*

In general, biomineralization starts with the deposition of amorphous minerals. The final crystalline minerals derive from the amorphous precursors (Weiner et al. 2009). In addition, biomineralization is, at least partially, enzyme catalyzed. This has been shown for all major classes of biominerals, both silica (synthesis of amorphous biosilica by the sponge enzyme silicatein; Krasko et al. 2000; Schröder et al. 2012; for a review, see Schröder et al. 2016), calcium carbonate (enzyme: carbonic anhydrase; see Wang et al. 2014f), as well as calcium phosphate/bone hydroxyapatite (enzyme: alkaline phosphatase [ALP]; see Müller et al. 2019a). It has been demonstrated that bone HA formation is initiated by the deposition of amorphous Ca-carbonate (ACC) (most likely catalyzed by the membrane-associated carbonic anhydrase IX), which is subsequently transformed to amorphous Ca-phosphate (ACP) and finally HA crystals (Wang et al. 2018). The phosphate is delivered by polyP (Müller et al. 2015a), in addition to phosphate from β-glycerophosphate, which are both enzymatically cleaved by ALP. Only the amorphous, ACC and ACP, but not the crystalline minerals involved in bone mineral formation (Tolba et al. 2016), as well as amorphous silica (Wang et al. 2014a) exhibit morphogenetic, osteoblastic activity, both in vitro and in vivo (for reviews, see Wang et al. 2014e, 2018).

In our studies, confirmed in later studies (Lotsari et al. 2018), strong evidence has been provided that HA formation comprises three phases: (1) Deposition of ACC "bio-seeds" (enzyme-catalyzed); (2) transformation of ACC to ACP (partially enzyme catalyzed); and finally, (3) maturation of ACP to crystalline HA (nonenzymatically) (Wang et al. 2018). The phosphate required for the carbonate-phosphate exchange reaction from ACC to ACP (Müller et al. 2015a) originates from polyP, which is enzymatically hydrolyzed by ALP. Both amorphous calcium salt minerals, ACC (Tolba et al. 2016; Wang et al. 2016b) and ACP (Müller et al. 2016d) can be stabilized by polyP. In the apatite crystal lattice of the finally deposited bone mineral a significant amount of carbonate (CO_3^{2-}) is found (about 5% [w/w]), which substitutes for phosphate or hydroxide ions (Tolba et al. 2016), supporting the proposition that bone Ca-phosphate deposition is preceded by

Ca-carbonate "bio-seed" formation. Due to this fact, as well as the presence of small amounts of Na^{+}, Mg^{2+}, and HPO_4^{2-} ions, the bone mineral has a nonstoichiometric composition of Ca^{2+} and PO_4^{3-} that does not correspond to the formula $Ca_{10}(PO_4)_6(OH)_2$ of bone HA (Von Euw et al. 2019).

PolyP plays a critical role in these processes, not only as an extracellular donor of metabolic energy, but also by the delivery of the orthophosphate required for the mineralization process, which is produced by the enzymatic, ALP-mediated hydrolytic breakdown of the polymer.

5.3 Conventional Scaffold/Implant Materials and Need for Improvement

A scaffold that can be applied as an implant material for tissue repair should not only be able to become perfectly integrated into the defect region with the surrounding uninjured tissue. It should also induce regeneration of the host tissue, which should ultimately lead to a replacement of the implant by native, functionally active tissue of the recipient. The regenerative capacity of different tissues shows great differences (Fig. 5.1). Among the osteoarticular tissues, cartilage, in particular, is characterized by a low capacity for regeneration/repair. The prerequisite for a regenerative active scaffold is that the scaffold material is able to trigger the cellular signaling pathways to induce the differentiation and proliferation of stem cells to restore the functionally active tissue in the defect area (Rice et al. 2013). Such regeneratively active scaffolds, which promote the growth and maturation of stem cells and are even suitable for the 3D printing of personalized, cell-laden scaffolds, could only be developed more recently (Fig. 5.2).

Originally, materials used for scaffolds/implants, e.g., for bone defects, were based on metals or ceramics. Such materials as titanium and alloys or alumina and zirconia, as well as carbon-based material (carbon nanotubes; Aoki et al. 2020), are biologically inert and are not replaced by functionally active host tissue (for a review, see Battafarano et al. 2021); Fig. 5.2. Therefore, materials were later developed that are biodegradable and bioactive. These materials have a composition similar to that of bone mineral, such as α- and β-tricalcium phosphate (TCP) or bioactive glasses (Hench and Jones 2015). Some of these materials are even able to induce bone formation (Jeong et al. 2019). A significant advance has been the development of scaffolds/implants made from biodegradable polymers that form networks that have some morphological resemblance to the structure of the ECM. Such scaffold/implant materials, which consist of either natural or synthetic or both natural and synthetic polymer fiber networks, have to date found increasing research interest in tissue regeneration and repair, and some have also been introduced into the clinic. The beneficial, advantageous properties of these materials are biocompatibility, biodegradability, cell interaction, stimulation of specific cellular responses, and regenerative activity (for a review, see Reddy et al. 2021).

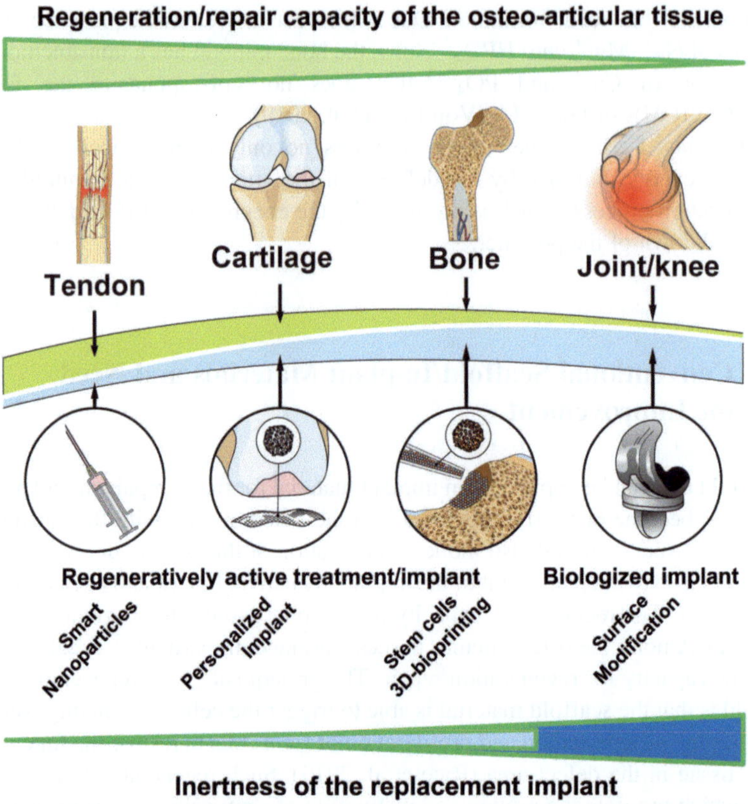

Fig. 5.1 Regeneration/repair capacities of various osteoarticular tissues and treatment/implant materials. In the past, mostly nonphysiological bio-inert implants were used to treat osteoarticular defects, which are now to be replaced by implants with regenerative potential. Such strategies require the availability of suitable regeneratively active biomaterials

The group of natural biopolymers comprises both proteins/peptides such as collagen (for a recent review, see Rezvani Ghomi et al. 2021) or silk fibroin (Neubauer et al. 2021), polysaccharides such as alginate, a copolymer consisting of blocks of (1→4)-linked β-D-mannuronate and α-L-guluronate units that can be hardened in the presence of calcium ions (Zarrintaj et al. 2018), hyaluronic acid, a polysaccharide composed of *N*-acetyl-glucosamide and D-glucuronic acid (Fallacara et al. 2018), chitosan, a copolymer of *N*-glucosamine and *N*-acetyl-glucosamine (Hu et al. 2020) or chondroitin sulfate, composed of glucuronic acid and *N*-acetyl-D-galactosamine (Ji et al. 2020), and other polymers such as poly(3-hydroxybutyrate), a polyhydroxyalkanoate polyester produced by microorganisms (Chai et al. 2020), which has been shown to form membrane channels together with polyP (Reusch 2014); Fig. 5.2. Collagen is the mostly used biopolymer because of its low immunogenicity and the presence of integrin binding sites (RGD motif), which promotes

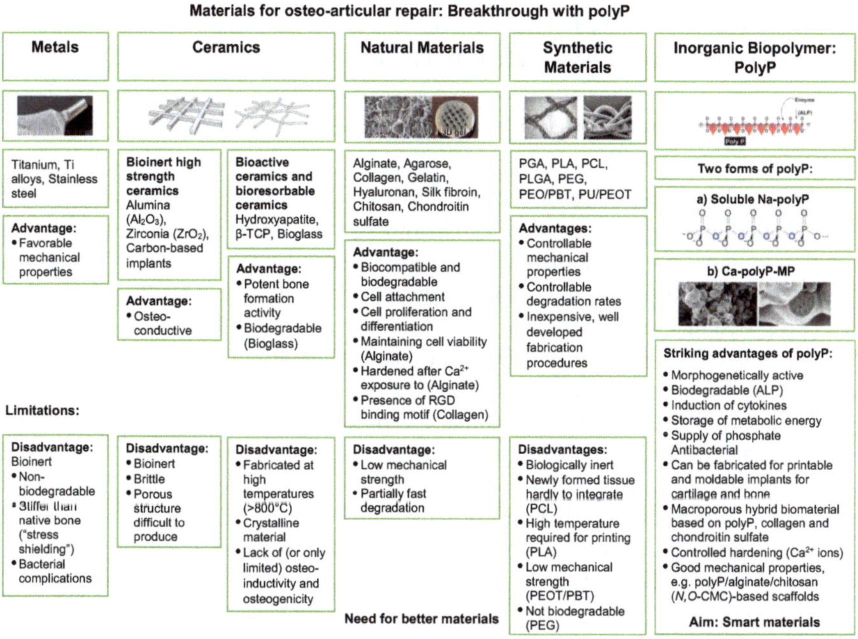

Fig. 5.2 Limitations (advances versus disadvantages) of conventional materials for osteoarticular repair in comparison to polyP. Abbreviations: HA, hyaluronic acid (hyaluronan); *N,O*-CMC, *N,O*-carboxymethyl chitosan; PBT, polybutylene terephthalate; PCL, poly(caprolactone); PEG, poly (ethylene glycol); PEO, polyethylene oxide; PEOT, polyethyleneoxide terephthalate; PGA, polyglycolic acid; PLA, polylactic acid; PLGA, poly(lactic-*co*-glycolic acid); PU, polyurethane; β-TCP, β-tricalcium phosphate

cell migration, adhesion, proliferation, and differentiation (Glowacki and Mizuno 2008; Liu et al. 2019).

Examples for synthetic polymers are poly(glycolic acid) (PGA), poly(lactic acid) (PLA), poly(ε-caprolactone) (PCL), poly(lactic-*co*-glycolic acid) (PLGA), and poly (ethylene glycol) (PEG), as well as copolymers like polyethylene oxide (PEO)/ polybutylene terephthalate (PBT) or polyurethane (PU)/polyethyleneoxide terephthalate (PEOT) (reviewed in: Battafarano et al. 2021; Reddy et al. 2021).

To some extent, the biopolymer networks formed by these materials, natural or synthetic, mimic the supramolecular architecture of the ECM. They exhibit good biocompatibility and (partially) good biodegradability, but only a few of them like collagen and fibronectin exhibit morphogenetic activity (Boes et al. 2009). In particular, synthetic polymers are bio-inert and must be supplemented with the growth factors needed by the cells for application in tissue engineering and regenerative medicine. Therefore, there is still a need for materials with better tissue regeneration/repair properties.

5.4 Inorganic Polyphosphate (PolyP), a Biomimetic Material for Tissue Regeneration

5.4.1 Occurrence and Chemistry of PolyP

Inorganic polyphosphates (polyP) are mainly linear polymers, formed by condensation of monomeric orthophosphate (P_i) (Kornberg et al. 1999; Kulaev et al. 2004). In addition to linear polymers, there are cyclic molecules, which are usually smaller in size (mostly 3, 4, or 6 P_i units), although the existence of large cyclic molecules has been proposed that might have played a role in the origin of life (Glonek 2021). The phosphoanhydride (P-O-P) bonds, which connect the tetrahedrally coordinated P_i units via a common oxygen atom, are energy-rich bonds (Müller et al. 2019a).

The occurrence of polyP in animals, especially humans, has been detected relatively late, (Lorenz et al. 1997; Leyhausen et al. 1998; Kornberg 1999; Schröder and Müller 1999), decades after the first description in yeasts and bacteria (Kulaev et al. 2004; Rao et al. 2009). Therefore, the study of the physiological function of polyP in humans, particularly its role as a potential target in the therapy or prevention of human diseases, began very late.

In humans, polyP appears to be present in almost all cells and tissues, both intra- and extracellularly. Particularly high concentrations of polyP are found in the platelets and in mast cells as well as in blood serum (Ruiz et al. 2004). The polyP concentrations are highest in the places of their intracellular formation and storage, in the mitochondria and even more so in the acidocalcisomes. In particular, the platelet acidocalcisomes, the dense granules, are assigned an important function as polyP storage organelles that accumulate large amounts of the polymer, which is released after platelet activation, e.g., at sites of tissue damage such as bone fractures. In fact, polyP has an important function in regulating and as an energy donor and phosphate source of bone mineralization, the formation of hydroxyapatite (HA), after hydrolysis to monomeric P_i (Omelon et al. 2009). Consequently, bone tissue is one of the human tissues with the highest polyP content (Leyhausen et al. 1998).

Physiologically, in humans, polyP is mostly present in the chain length range from 50 to 100 P_i units, e.g., as polyP in extracellular body fluids released from platelets after activation (Morrissey et al. 2012), although longer polyP chains can also be detected, e.g., in nanoparticles found on the platelet surface (Verhoef et al. 2017). At neutral pH, polyP is a strong polyanion with a negative charge at each internal P_i unit and two negative charges at the terminal P_i units. The two dissociation constants are $pK_1 = 2.2$ (both internal and terminal) and $pK_2 = 7.2$ (only terminal) (Kirk-Othmer Encyclopedia of Chemical Technology 1991). If present as a sodium salt (Na-polyP) with chain lengths of less than 100 P_i units, polyP is readily soluble in water (van Wazer 1958). The solubility of the polymer, however, significantly decreases in the presence, especially of excess amounts of divalent cations, such as Ca^{2+}, Mg^{2+}, Fe^{2+}, Mn^{2+}, and Zn^{2+}.

At room temperature, in neutral or alkaline solutions, polyP is stable over long periods of time (van Wazer 1958; Kulaev et al. 2004). This behavior is opposed to the thermodynamic properties of the polymer. Under standard conditions, the thermodynamic equilibrium of polyP in aqueous solutions is almost completely on the side of the P_i monomers. The standard Gibbs free energy (ΔG^0) for hydrolysis of each phosphoanhydride bond in the linear polyP equals those liberated during hydrolysis of the α-β and β-γ P-O-P bonds in ATP (-30.5 kJ mol^{-1}). Presumably, the clustered multiple negative charges of the surface of the polyanionic polyP impede with the hydrolytic attack of water molecules (Thilo and Wicker 1957; Westheimer 1987). As a result, the hydrolytic cleavage reaction is associated with large activation energy (E_a). In the presence of H^+ ions, E_a is markedly reduced and hydrolysis readily occurs (De Jager and Heyns 1998a, b).

5.4.2 Physiological Functions of PolyP

In recent years, research on polyP in mammalian cells and tissues, including humans, has identified several functions of the polymer. PolyP has been shown to be involved in bone formation by inducing the mineralization of bone-forming cells, as demonstrated both in vitro and in vivo (Leyhausen et al. 1998; Schröder et al. 2000; Müller et al. 2011; Wang et al. 2012, 2013b). Administered as a complex with Ca^{2+} ions, polyP strongly stimulates HA formation of osteoblast-like, mineralization-competent SaOS-2 cells (Wang et al. 2013a), parallel to an increased gene expression of the cytokine bone morphogenetic protein 2 (BMP2), an inducer of osteoblast differentiation (Wang et al. 2013a). In addition, polyP enhances the expression of a series of further proteins involved in bone formation, such as ALP. Moreover, polyP acts as a source of P_i needed for HA formation (Omelon and Grynpas 2008; Omelon et al. 2009; Müller et al. 2015a).

Another important property of polyP, in particular with respect to its potential application in bone regeneration/repair and wound healing, is its stimulatory effect on microvascularization (Müller et al. 2018b). In addition, polyP has an important function in the regulation of blood coagulation and fibrinolysis, which has been extensively studied (Smith et al. 2006; Müller et al. 2009). PolyP has also been reported to be involved in the control of the complement system (Wijeyewickrema et al. 2016) and to act as an activator of the mitochondrial permeability transition pore (Seidlmayer et al. 2012) and of mTOR kinase (Wang et al. 2003).

The function of polyP as an energy storage and energy supply especially in the extracellular space (extracellular matrix) (Müller et al. 2015c, 2017g; Wang et al. 2016d) is described in Sect. 5.7.2.

5.4.3 Metabolism of PolyP

In human and other mammalian cells, polyP is synthesized in the mitochondria (Pavlov et al. 2010; Angelova et al. 2016) and most likely also by the acidocalcisomes, which are the main storage organelles of the polymer (reviewed in: Docampo et al. 2005). In the mitochondria, the polyP synthesis is driven by the energy of a membrane gradient/potential; it is suppressed by the mitochondrial ATP synthase inhibitor oligomycin (Pavlov et al. 2010). The acidocalcisomal synthesis of polyP is probably mediated, like in yeast, by the vacuolar transporter chaperone (Vtc) complex, which is located at the acidocalcisome membrane (Hothorn et al. 2009). In the yeast Vtc, a polyP polymerase activity, which synthesizes polyP from ATP, has been identified at the cytoplasmic side of the complex. The reaction catalyzed by this enzyme involves both the cleavage of an energy-rich phosphoanhydride bond (the β-γ phosphoanhydride bond in ATP) and the formation of a new phosphoanhydride bond (in the growing polyP chain); i.e., the ΔG^0 of the overall reaction is close to zero. Therefore, it seems important that the catalytic reaction (polyP synthesis on the cytosolic side of the Vtc complex) is coupled with the transport and accumulation of the polymer into the acidocalcisome in order to remove the product from the chemical equilibrium.

The acidocalcisomes are found in close proximity to the mitochondria, the organelles responsible for ATP synthesis via the F_OF_1–ATP synthase complex of the respiratory chain. Although the mitochondria themselves are able to utilize the ATP for polyP synthesis (Abramov et al. 2007; Pavlov et al. 2010), this mitochondrial polyP most likely cannot leave these organelles to become available in the extramitochondrial space. The mitochondria are only able to export the ATP, which is formed by the F_OF_1–ATP synthase, via an ADP/ATP antiporter, the adenine nucleotide translocase (ANT), which is located in the mitochondrial inner membrane, from the mitochondrial matrix to the intermembrane space. From there the ATP is released into the extramitochondrial cytosol via the voltage-dependent anion channels (VDAC), which are present in the mitochondrial outer membrane (Fig. 5.3a). Facilitated by the close proximity of the acidocalcisome, the ATP leaving the VDAC has been proposed to serve as a substrate for the Vtc polyP polymerase (Fig. 5.3b), resulting in the synthesis and import of polyP in the acidocalcisome (Müller et al. 2019a).

It is supposed that the flux of ATP from the mitochondria to the acidocalcisomes is critical for the rate of polyP synthesis by the acidocalcisomal Vtc complex. While the ADP/ATP exchange rate of the ANT in the mitochondrial inner membrane is controlled by the mitochondrial membrane potential (Chinopoulos et al. 2009), the ATP flux across the outer mitochondrial membrane via the VDAC complex has been reported to be regulated by tubulin which can close the channel (Fig. 5.3c). Only the detyrosinated but not the tyrosinated tubulin is able to close the channel (Sheldon et al. 2015). The enzyme that catalyzes the posttranslational tyrosination at the carboxyl end of the α-subunit of the tubulin dimer is a specific tubulin-tyrosine ligase that depends on the presence of ATP (Schröder et al. 1985). It is reasonable to

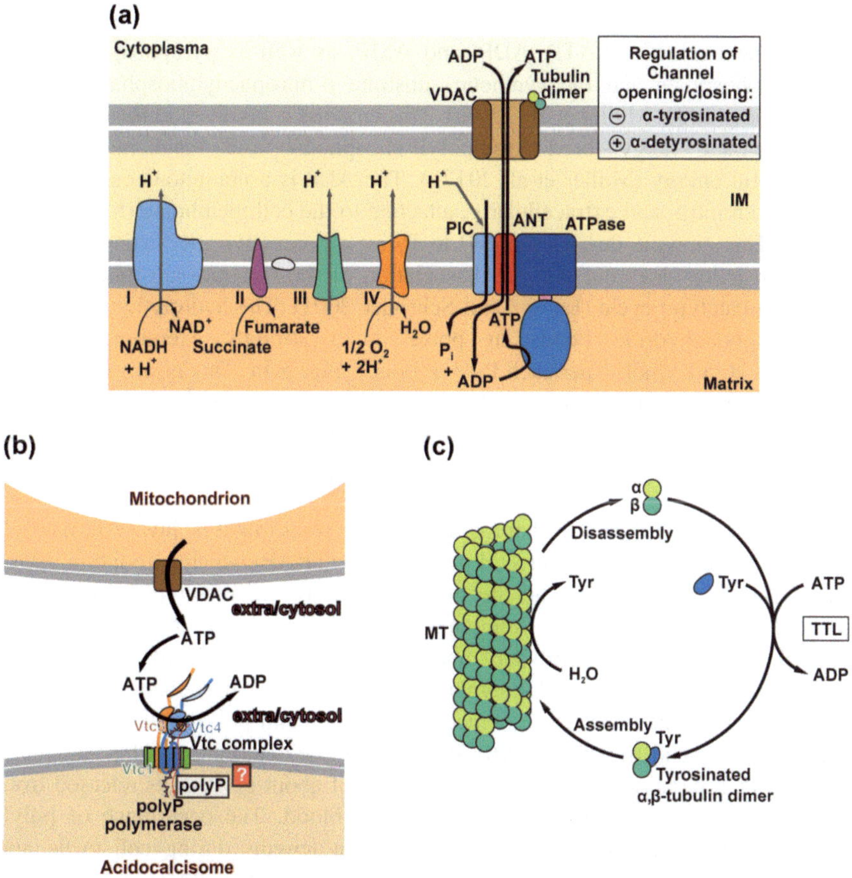

Fig. 5.3 Proposed transfer of mitochondrially produced ATP to the acidocalcisomal Vtc complex and control. (**a**) ATP produced by the mitochondrial F_OF_1–ATP synthase (ATPase) complex, driven by the H^+ gradient at the inner mitochondrial membrane, is translocated from the mitochondrial matrix via the ANT into the intermembrane space. The export of ATP, from there, into the cytosol across the VDAC in the outer mitochondrial membrane is regulated by tubulin. Tyrosinated α-tubulin closes the channel, while detyrosinated α-tubulin opens the channel. (**b**) The exported mitochondrial ATP is probably used by the polyP polymerase activity of the acidocalcisomes for synthesis of polyP, which is translocated into the lumen of the acidocalcisomes. Adapted with permission from Müller et al. (2021a). Copyright 2021, Royal Society of Chemistry. (**c**) The tyrosination of the αβ-tubulin dimer is an ATP-dependent process, catalyzed by tubulin-tyrosine-ligase (TTL). It is suggested that at high levels of cytosolic ATP tubulin is tyrosinated and both ATP export by VDAC and polyP accumulation in the acidocalcisomes are downregulated, while at low cytosolic ATP levels an upregulation occurs

assume that this enzyme is involved in controlling the rate of Vtc-mediated polyP formation by disrupting the flow of ATP through the VDAC channel at high levels of cytosolic ATP.

The major polyP-degrading enzyme in humans is the ALP, which also degrades the adenine nucleotides ATP, ADP, and AMP, as well as pyrophosphate (PP_i), β-glycerophosphate and the synthetic substrate *p*-nitrophenylphosphate (Lorenz and Schröder 2001; Müller et al. 2019a). This enzyme is involved in the conversion of the chemical energy stored in the polyP phosphoanhydride bonds into metabolically useful energy (Müller et al. 2017g). The ALP is a ubiquitous enzyme. It is present both intra- and extracellularly, attached to the cell membrane (Müller et al. 2019a). This enzyme degrades polyP to P_i by a processive mechanism, i.e., the enzyme reaction does not involve dissociation of the enzyme–substrate complex after each catalytic cycle (Lorenz and Schröder 2001). The hydrolytic cleavage of the phosphoanhydride bonds in polyP is a strongly exergonic reaction ($\Delta G^0 = -30.5$ kJ mol^{-1} per each P–O–P bond); Fig. 5.4a. This means that a large amount of energy can be produced by degradation of a single polyP molecule in a continuous reaction by a single enzyme molecule; e.g., 39×-30.5 kJ $mol^{-1} = -1189.5$ kJ mol^{-1} by complete hydrolysis of $polyP_{40}$ (chain length of 40 P_i units).

In addition to ALP, two further mammalian enzymes might be involved in polyP catabolism; the tartrate-resistant acid phosphatase (TRAP) which is found in osteoclasts (Harada et al. 2013) and the human metastasis regulator protein H-prune (Tammenkoski et al. 2008). Both enzymes hydrolyze only short-chain polyP, while the ALP degrades both short- and long-chain polyP molecules (Lorenz and Schröder 2001). Also, an endopolyphosphatase has been described and partially purified from mammalian tissue, which splits long-chain polyP into $polyP_{60}$ fragments (Kumble and Kornberg 1996). It remains to be determined if this enzyme is involved in generation of the short-chain polyP of about 60 P_i units released from activated platelets into extracellular fluids like blood. The occurrence of polyP fractions with a sharply limited range of chain lengths during cell cycle and apoptosis has also been detected in HL-60 cells (Lorenz et al. 1997).

5.5 PolyP Nano/Microparticles

5.5.1 Physiological Deposits

Physiologically, polyP in humans is present both in a soluble and in a particulate form, as polyP nano/microparticles. The major cellular deposits of polyP nano/microparticles are found in the acidocalcisomes (Ruiz et al. 2004; Docampo and Huang 2016). These organelles correspond to the bacterial volutin granules. Large amounts of acidocalcisomes ("dense granules") are found in the blood platelets (Docampo et al. 2013). The concentration of polyP in the acidocalcisomes can amount to up to 130 mM (in terms of P_i units), and the polyP level of the platelets, due to their high content of dense granules, is about 1.1 mM (Wang et al. 2016d, 2018). This value exceeds the polyP level in other mammalian cells by 10–20-fold (see Kumble and Kornberg 1995). The accumulation of polyP in the

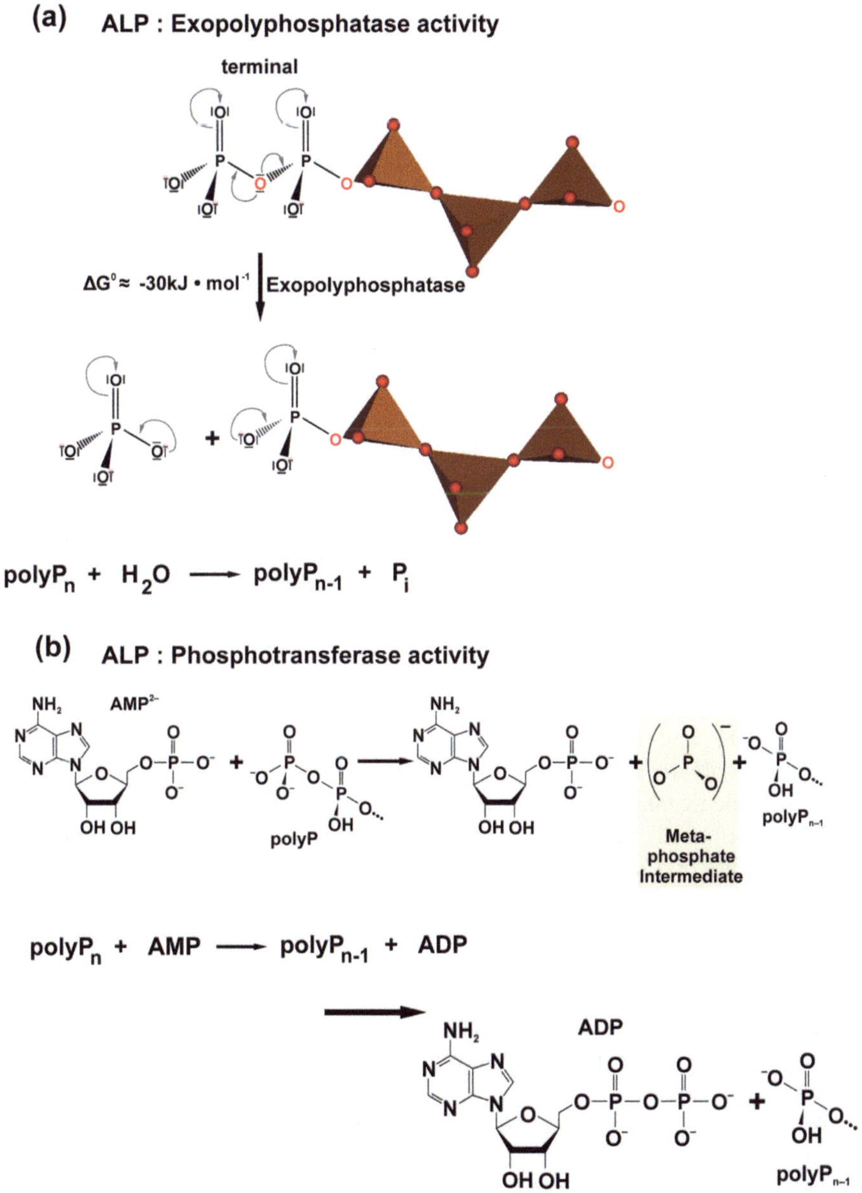

Fig. 5.4 Enzymatic activities of alkaline phosphatase. (**a**) Exopolyphosphatase reaction, resulting in a hydrolytic cleavage of the terminal phosphoanhydride bond of the polyP molecule. (**b**) Phosphotransferase reaction. The transfer of the phosphate from polyP to AMP most likely follows a dissociative mechanism, proceeding via formation of a metaphosphate intermediate. Adapted with permission from ref. Müller et al. 2019a. Copyright 2019, American Chemical Society

acidocalcisomes mediated by Vtc complex has been studied in detail in yeast and Protozoa (Lander et al. 2013; Ulrich et al. 2014).

The polyP enriched in the platelets is released after activation of the cells (Ruiz et al. 2004; Müller et al. 2009; Moreno-Sanchez et al. 2012). The chain length of the polyP molecules enriched in the dense granules and secreted from the platelets have been reported to range between 60 and 100 P_i units (Ruiz et al. 2004; Müller et al. 2009). However, on the platelet surface, nanoparticles of long-chain polyP salts have been detected that are retained on the surface after platelet activation (Verhoef et al. 2017; Labberton et al. 2018).

5.5.2 *Bioinspired Synthesis*

Crystalline phosphate particles conventionally used in bone tissue engineering show little or no biological activity, such as crystalline HA, in contrast to amorphous polyP particles, which are more biocompatible and morphogenetically more active. Therefore, in a bioinspired way, we synthesized amorphous nano/microparticles of polyP to mimic the polyP deposits that are found in acidocalcisomes. Nano/microparticles of various metal salts of polyP have been prepared, starting from soluble polyP (Na-polyP) and various counterions, comprising both divalent cations, such as Ca^{2+}, Mg^{2+}, Sr^{2+}, and Zn^{2+} (Müller et al. 2015g, 2017e, 2018c, 2020b), or trivalent cations, such as Gd^{3+} (Wang et al. 2016c). Usually, polyP of an average chain length of around 40 P_i units ($polyP_{40}$) is used. The synthetic procedure requires an alkaline pH (usually pH 10 is applied) and an excess amount of the counterion, i.e., a substoichiometric ratio of polyP (based on P_i) to the salt forming metal ion; routinely a phosphorus:metal ratio of 1:2 is used. The sizes of the synthesized spherical particles, e.g., of Ca-polyP nano/microparticles, usually vary between 80 and 200 nm (Müller et al. 2015g). They can be adjusted to larger or smaller dimensions by varying the preparation protocol (change of the drop speed during mixing of the components or application of ultrafiltration procedures). The particles have a porous internal structure and are stable in the dried state over long periods of time or in protein-free aqueous solutions at neutral or alkaline pH. Their amorphous character has been confirmed by powder X-ray diffraction (XRD) analysis (Müller et al. 2015g, 2016a, 2017e).

As shown in Fig. 5.5, the formation of the nanoparticles, e.g., Ca-polyP nanoparticles, seems to involve a rolling up of the polyP chains. At a stoichiometric excess of Ca^{2+} ions, compared to phosphate, almost all binding sites of the polyanionic polymer for divalent counterions are occupied by Ca^{2+}. It is assumed that the polyP chain thereby becomes twisted and coiled under the formation of particulate structure. Numerous polyP molecules will be necessary to form a 20-nm-sized polyP nanoparticle. Assuming a P–O distance of 1.61 Å and a four-phosphate-per-turn helix structure of the polyP chain, resulting in a length reduction by 38% (Glonek 2021), a longitudinal size of about 110 Å (1.1 nm) for a $polyP_{40}$ molecule can be calculated. Also, considering the space of the calcium ions (ionic radius

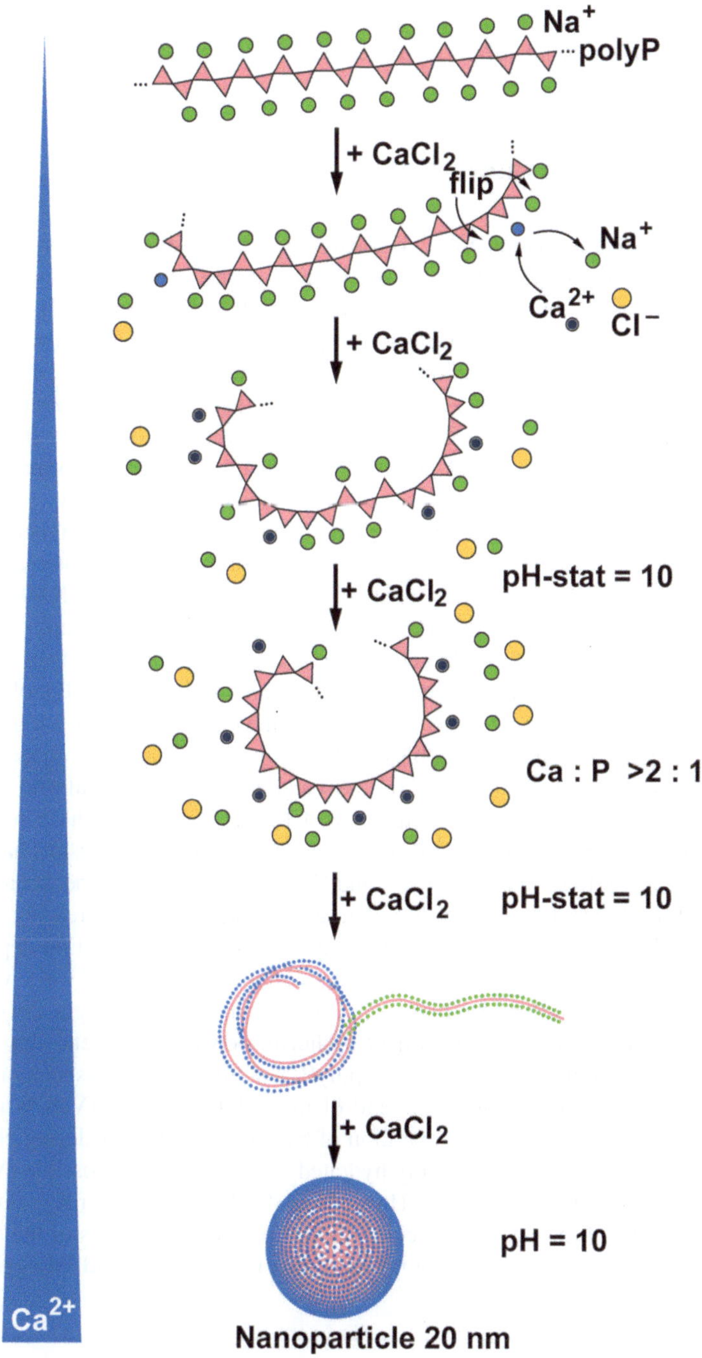

Fig. 5.5 Formation of amorphous Ca-polyP-nanoparticles at alkaline pH and a superstoichiometric Ca^{2+} : phosphate ratio. The Ca^{2+} concentration is gradually increased to a final Ca^{2+} : phosphate ratio of 2 : 1 under pH-stat conditions (pH 10). The occupation of almost all binding sites for divalent cations along the polyanion leads to a rolling up of the polyP chain, which forms a

0.1 nm), it can be estimated that a 20-nm Ca-polyP nanoparticle could be composed of up to several thousands of polyP molecules.

5.6 PolyP Coacervates

In contrast to the amorphous polyP nanoparticles which are formed from Na-polyP at alkaline conditions (pH 10) in the presence of excess amounts of metal ions, a polyP coacervate is obtained from the soluble polyP at neutral pH (pH 7), and an approximately stoichiometric ratio between phosphate and metal ion (Müller et al. 2018d). This process involves a liquid–liquid phase separation (Yewdall et al. 2021), which leads to a more viscous, colloid-rich phase that mainly contains the longer polyP molecules, and to a less viscous phase with only the residual very short polyP molecules (Franco et al. 2020). A possible mechanism of coacervate formation from Na-polyP during addition of increasing concentrations of Ca^{2+} is shown in Fig. 5.6. The addition of Ca^{2+} leads to ionic interaction and complex formation between the divalent cation and the negatively charged oxygens of two adjacent internal P_i units of the polyP chain. Based on the results of Eu^{3+} luminescence, ^{31}P NMR, and IR studies, it has been proposed that Ca^{2+} successively binds to two groups of binding sites in the polyP chain (Dias Filho et al. 2005). The first group consists of cage-like sites with the encapsulated divalent cation, which are used at low Ca:P molar ratios. The second group is used after the occupation of the cage-like sites at a higher Ca:P ratio. Binding to the latter sites enables the Ca^{2+} ions to interact with adjacent polyP chains by forming cross-links (Ca^{2+} bridges) and with water molecules to adopt an octahedral geometry (coordination number of 6). As a result, a phase separation occurs and a polymer-rich Ca-polyP coacervate is formed. Similar results have been reported for other divalent cations (Ni^{2+} and Co^{2+}) at a metal:P ratio between 0.1 and 2 (Silva et al. 2008). It has been described that the addition of increasing concentrations of divalent cations to soluble Na-polyP leads to a concentration-dependent, almost constant decrease in the pH of the solution (Frankenthal 1944), which is accelerated at a metal/P molar ratio of $\sim$0.18 (Momeni and Filiaggi 2014). This effect is explained by the formation of a chelate between the metal ion and the terminal and penultimate P_i unit of the polyP, resulting in the dissociation of the second, less acidic OH group at the end of the polyP molecule (Van Wazer and Campanella 1950), but also the formation of hydroxyl complexes during chelation by hydrolysis of the chelated partially hydrated divalent cation is possible (Momeni and Filiaggi 2014). The drop of the pH value and the decrease in the pK_a value of the terminal P_i units of the polyP molecules depend on the ionic radius of the divalent cation and increase with decreasing radius of this ion (Momeni and Filiaggi 2014).

Fig. 5.5 (continued) particulate structure. Due to the small size of a polyP molecule with a maximum length of approx. 1 nm (for $polyP_{40}$), which decreases during formation of the coiled structure, thousands of polyP molecules are required to build-up a 20-nm Ca-polyP nanoparticle

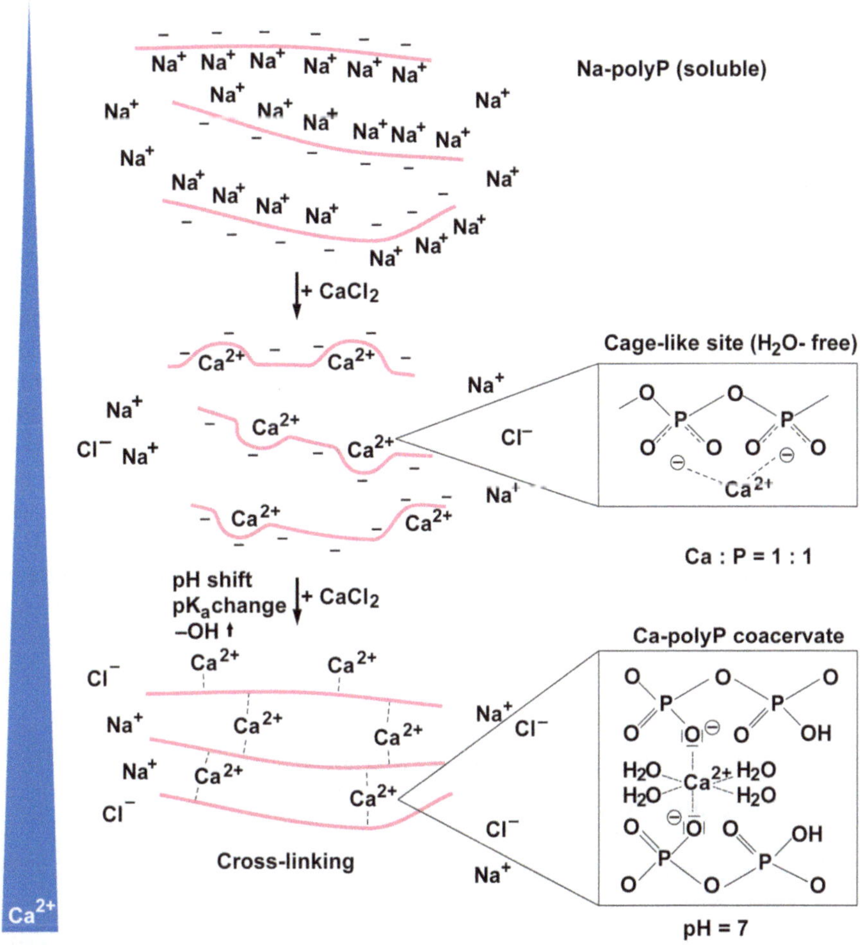

Fig. 5.6 Formation of a coacervate from water-soluble Na-polyP caused by gradually increasing the Ca^{2+} ion concentration. With a low Ca^{2+}: phosphate ratio, it is assumed that the Ca^{2+} ions are bound in water-free or almost water-free cage-like structures, which are formed by wrapping the polyP chain around the Ca^{2+} ions, which reach a coordination number of 6. A further increase in the Ca^{2+} concentration (up to a stoichiometric Ca^{2+}: phosphate ratio of 1 : 1) leads to a concentration-dependent decrease in pH of the solution, most likely caused by the formation of Ca^{2+} chelates with two terminal P_i units at the end of the polyP molecule, which leads to the dissociation of the second weakly acidic hydrogen atom of the terminal P_i unit, which then becomes strongly acidic. Due to the decrease in the pH value and the increasing number of non-dissociated internal P_i units, the polyP structure then rearranges with phase separation and coacervate formation, caused by the cross-linking of the polyP chains via Ca^{2+} bridge formation between individual, negatively charged P_i units present in the polymers

As shown in Fig. 5.6, the decrease in pH of the Na-polyP solution during the stepwise addition of Ca^{2+} ions leads to an increasing number of non-dissociated OH groups at the polyP chain. As a result, the number of possible intra-chain Ca^{2+} chelate complexes decreases and the likelihood of Ca^{2+} bridge formation between two opposing polyP molecules increases.

A coacervate can also be formed from the amorphous Ca-polyP nanoparticles, which have the propensity to transform into a coacervate after exposure to protein-containing fluids. This process is accompanied by a reduction of the strongly negative zeta (ζ) potential of the particles (−33.6 mV; Müller et al. 2018d). This ζ potential prevents the aggregation of the particles if suspended in aqueous solution.

In contrast to the amorphous polyP particles, which are characterized by high stability but relatively low biocompatibility, the polyP coacervate phase is less stable, but represents the biologically (morphogenetically) active form of the polymer (Müller et al. 2018d). The polyP coacervate forms an adaptable matrix that supports cell ingrowth as well as the growth and differentiation of cells embedded in the coacervate (Müller et al. 2018d), mimicking a stem cell niche (Moore and Lemischka 2006). Even more, and not shown by any other material, polyP also provides the cells with the required metabolic energy (Müller et al. 2019a). This unique property of polyP is of advantage in tissue engineering, especially in bone or cartilage tissue engineering, where high cell densities are needed, which is a challenging task due to the limited nutrient supply.

5.7 Basic Properties of PolyP as a Regeneratively Active Inorganic Polymer

5.7.1 Morphogenetic Activity

A prominent property of the polyP nano/microparticles is their morphogenetic activity. The particles are able to induce the differentiation of stem cells via specific gene induction (for reviews, see Wang et al. 2014c, d, 2016d; Müller et al. 2015e). By choosing the suitable cationic counterion of the polyanionic polyP, it is even possible to control the direction of differentiation. Mesenchymal stem cells (MSC) can be directed either into the osteogenic, chondrogenic, myogenic, or tenogenic lineage (Fig. 5.7). This is achieved by controlling the expression of those genes which are involved in decisive steps in differentiation (Müller et al. 2015d, e, 2017d, e; Wang et al. 2016d, 2018). For example, Ca- or Sr-polyP nano/microparticles cause the regeneration of mineralized bone tissue (Müller et al. 2017e), while Mg-polyP nano/microparticles promote the regeneration and repair of cartilage (Müller et al. 2016a). In this property, the polyP particles differ from most other materials used in tissue engineering and repair. These materials are mostly biologically inactive and require additives like cytokines for use as implants, to induce stem cell maturation to the functionally active cells. PolyP, on the other hand, is able to

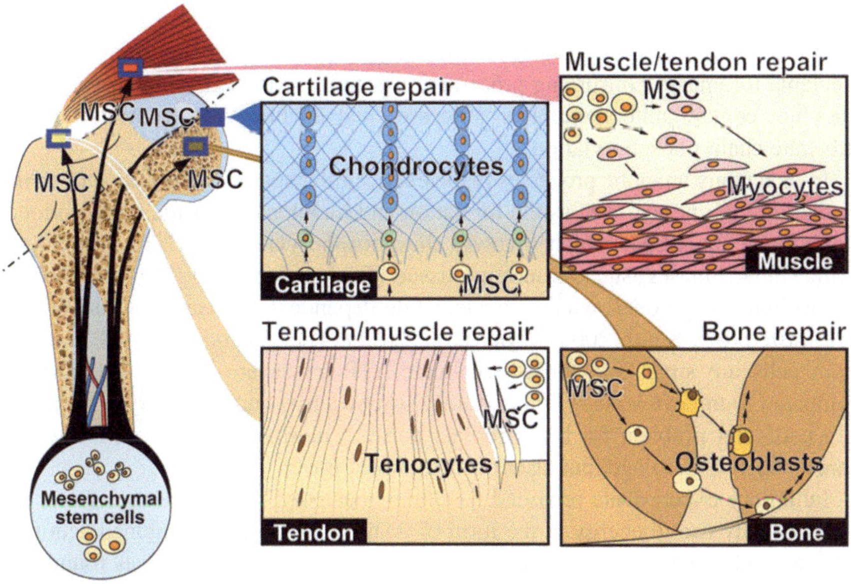

Fig. 5.7 Control of the direction of differentiation induced by morphogenetically active amorphous polyP nano/microparticles. Mesenchymal stem cells (MSC) can be directed either into the chondrogenic, myogenic, tenogenic or osteogenic lineage by choosing the suitable counterion of the polyanionic polyP; preferential differentiation of MSC into chondrocytes by Mg-polyP or Ca-polyP, into myocytes by Zn-polyP, into tenocytes by Mg-polyP, and into osteoblasts by Ca-polyP or Sr-polyP

induce the host cells to synthesize these cytokines themselves (Müller et al. 2015d, 2017e). This property makes polyP-based materials ideally suited for regeneration/repair from bone and cartilage defects to wound healing.

In addition, the type of the divalent counterion also influences the mechanical behavior of the polyP-based material (Müller et al. 2015d; Ackermann et al. 2018). With amorphous Mg-polyP, for example, materials can be obtained that show viscoelastic properties similar to cartilage. The divalent cations can also be used to harden polyP-based materials during 3D printing. For example, it is possible to print hydrogels from polyP in combination with biologically inert polymers (e.g., chitosan derivatives or alginate) in a curing solution with Ca^{2+} ions. During printing into the Ca^{2+} solution, the material hardens, enabling the fabrication of personalized scaffolds/implants.

5.7.2 *Generation of Metabolic Energy*

PolyP molecules, which can consist of up to dozen high-energy phosphoanhydride bonds, contain large amounts of metabolically usable energy, a multiple of the energy stored in ATP (see Sect. 5.4.3). No other biomolecule can store so much chemical energy in its energy-rich bonds. But how can this chemical energy be made available for energy consuming processes, especially in the extracellular space that does not contain mitochondria or processes that enable ATP synthesis through substrate chain phosphorylation?

In fact, many anabolic processes in the extracellular space strongly depend on the availability of energy, such as the formation and maintenance of the organizational structure of the ECM, which involves the phosphorylation of ECM proteins by extracellular kinases, such as in bone tissue (Tagliabracci et al. 2015), or certain extracellular chaperones, which are nucleotide-dependent, such as clusterin (Wilson and Easterbrook-Smith 2000; Wyatt et al. 2011). The quantities of ATP released from cells are small and the extracellular ATP levels are low (1–100 nM), if compared to the cytosolic ATP (1–10 mM) (Schwiebert and Zsembery 2003). This is a particular problem for bradytrophic, non-vascularized tissues with a low cell density such as cartilage, but also for mineralized bone.

Inhibitory experiments provided first hints that polyP can serve as a donor for metabolically useful energy in the form of ADP or ATP, via a combined action of ALP and adenylate kinase (ADK) (Müller et al. 2015c, 2017g). It was found that exposure of SaOS-2 cells to polyP causes an upregulation of extracellular ATP (Müller et al. 2015c). The increase in ATP level can be prevented by levamisole (ALP inhibitor), or by depletion of extracellular ATP with apyrase. More importantly, the application of the ADK inhibitor P^1,P^5-di(adenosine-5′) pentaphosphate (Ap_5A) causes a shift in the extracellular ATP/ADP ratio in favor of increased ADP levels (Müller et al. 2017g).

These results indicated that the ALP-catalyzed cleavage of the phosphoanhydride bonds in polyP can lead to the phosphorylation of AMP to ADP, and subsequent phosphorylation of ADP to ATP via the ADK reaction (Fig. 5.8). The latter enzyme catalyzes the interconversion of two ADP molecules into one AMP and one ATP. In other words, the formation of ATP from polyP is based on the concerted action of two enzyme reactions, catalyzed by ALP and ADK.

From the results, it was concluded that there must be a reactive phospho intermediate that enables the transfer of the terminal high-energy P_i from polyP to AMP. Based on theoretical considerations and the current state of knowledge about the mechanism of phosphotransfer reactions, the existence of an metaphosphate intermediate has been postulated, which enables ADP formation from polyP and the coupling of ALP and ADK reactions (Müller et al. 2019a).

5.7.3 *The Coupled ALP/ADK Reaction*

The ALP is a homodimeric metalloenzyme whereby each monomer is linked with the plasma membrane via a glycosyl-phosphatidylinositol (GPI) anchor (Hoylaerts et al. 1997; Millan 2006). Each catalytic site of the ALP dimer contains two Zn^{2+}

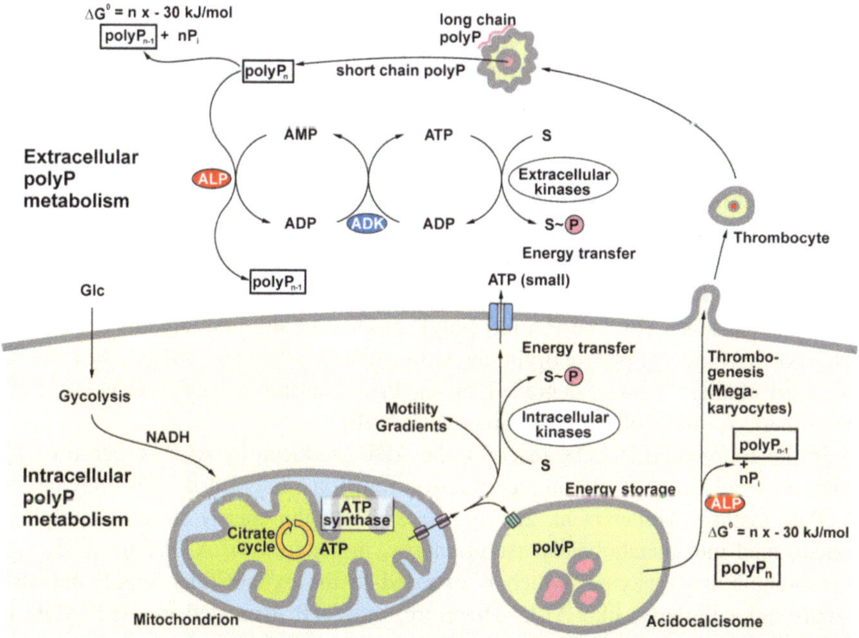

Fig. 5.8 Generation of ATP through the combined action of ALP and ADK. Extracellularly, polyP released from thrombocytes (platelets) is either hydrolyzed to orthophosphate by ALP with release of a huge amount of energy in the form of heat ($\Delta G^0 = -30\ \text{kJ mol}^{-1}$ per phosphoanhydride bond) or the energy-rich phosphate is transferred to AMP with formation of ADP. Two ADP molecules are then converted by ADK into AMP and ATP, which can be used by extracellular kinases. Intracellularly, polyP is formed in the mitochondria and probably the acidocalcisomes, which serve polyP storage organelles. ATP is synthesized by the mitochondrial ATP synthase and, after being released from these organelles, is used as a substrate for intracellular kinases or other energy consuming intracellular reactions or is exported to the extracellular space (only small amounts). The acidocalcisomal polyP is packaged in platelets after breakage of the thrombocyte precursor cells (megakaryocytes) or degraded intracellularly. Glc, glucose

ions and one Mg^{2+} ion, which are involved in the catalytic mechanism of the enzyme (Millan 2006).

Recently, a mechanism for the generation of ADP/ATP from polyP through the combined action of ALP and ADK has been proposed (Müller et al. 2019a). Of the three possible mechanisms for the ALP-mediated phosphoryl-transfer reaction (dissociative, associative, and concerted), the dissociative mechanism appears to be the most likely. This mechanism involves the formation of a metaphosphate species after cleavage of the terminal P_i group from polyP (Fig. 5.4b). The enzyme (ALP)-bound metaphosphate is then transferred to AMP, with the formation of ADP, which is then immediately taken up by the ADK to form AMP and ATP using a second ADP molecule.

In this reaction, Mg^{2+} ions, which are involved in the allosteric activation of ALP, are of critical importance. Mg^{2+} binds to the ALP dimer with negative cooperativity.

The Mg^{2+}-containing ALP subunit has a high affinity to polyP, while the ALP subunit lacking Mg^{2+} has a low affinity to the substrate. According to our model (Müller et al. 2019a), the affinity to the substrate of the two ALP subunits (low affinity and high affinity) changes alternatingly due to alternating conformational changes of the subunits during the reaction. In the first step, the Mg^{2+}-containing high-affinity ALP subunit binds polyP. The product of the enzyme reaction, ADP, formed after polyP cleavage and phosphoryl transfer to AMP via the metaphosphate intermediate, is only released from this subunit after a conformational change, which is induced by binding of Mg^{2+} to the second low-affinity subunit. Thereby, this subunit turns from low-affinity to high-affinity state that then binds, in the second step, the shortened (by one P_i unit) polyP product of the first subunit to perform the next cleavage and phosphoryl transfer reaction (Müller et al. 2019a). As a result, the polyP chain is stepwise degraded, by one P_i unit during each catalytic cycle of the monomeric subunits of the ALP dimer (Fig. 5.4b).

In the coupled ALP-ADK reaction, the ADP produced by ALP is then used by ADK as a substrate for the interconversion reaction catalyzed by this enzyme for ATP formation (Müller et al. 2017g). As a result, the energy stored in polyP is transformed into metabolically useful energy, in the form of ATP. The ADK is a membrane-bound enzyme, which is exposed to the extracellular space, but also occurs intracellularly, like ALP. Therefore, the ATP generated by ALP/ADK is available as a substrate for kinase reactions or other ATP-dependent processes both in the extracellular space and in the intracellular space (Müller et al. 2019a); Fig. 5.8.

5.8 PolyP Particles Used for Tissue Regeneration

5.8.1 Calcium-PolyP

The Ca^{2+} complexes of polyP as well as amorphous Ca-polyP nano/microparticles turned out to strongly accelerate bone HA formation and the healing of bone defects. Using osteoblast-like SaOS-2 cells, Ca-polyP was found not only to act as a potent inducer of expression of the main enzyme involved in mineralization, the ALP, but also to increase the activity of the enzyme (Müller et al. 2011). On the contrary, the maturation of osteoclasts and TRAP expression were found to be downregulated after exposure of RAW 264.7 cells to Ca-polyP (Wang et al. 2013a). PolyP also upregulates the expression of transcription factor RUNX2 (Runt-related transcription factor 2; a marker of osteoblastic differentiation) (Wang et al. 2014b) and induces an enhanced expression of BMP2 gene (Müller et al. 2011), in addition to other genes involved in bone formation, such as those of collagen type I, osteocalcin, osterix, and bone sialoprotein (Usui et al. 2010).

PolyP has also the potential to improve the reduced bone mineral density in osteoporotic patients. It was found that polyP induces the expression of osteoprotegerin (OPG) (Wang et al. 2012, 2013c). The reduced expression of this cytokine during menopause is associated with the pathogenesis of osteoporosis (Liu and Zhang 2015). Therefore, by inducing the expression of OPG, in addition to BMP2 and ALP, polyP might be of interest in prophylaxis and treatment of bone fractures caused by the disease (Wang et al. 2012, 2013c).

5.8.2 Magnesium-PolyP

As opposed to Ca-polyP nano/microparticles that induce RUNX2 expression, amorphous Mg-polyP nano/microparticles upregulate the expression of the transcription factor SOX9 (chondrogenic differentiation marker); in addition, these particles induce the expression of collagen types 2A1 and 3A1, ALP and aggrecan (Müller et al. 2016a, c; Wang et al. 2016a). Moreover, scaffolds containing Mg-polyP particles have been found to support the infiltration of the matrix by chondrocytes (Müller et al. 2016c).

5.8.3 Strontium-PolyP

The amorphous Sr-polyP nano/microparticles were found to enhance the growth and mineralization of SaOS-2 cells and MSC even more efficiently than the Ca-polyP particles (Müller et al. 2017e). In contrast to the Ca-polyP particles, the Sr-polyP particles strongly upregulate the expression of the osteocyte-specific *SOST* gene (Müller et al. 2017e). Sclerostin, the protein product of the *SOST* gene, is a Wnt antagonist and therefore acts as an inhibitor of the canonical Wnt/β-catenin signaling pathway (Malinauskas and Jones 2014). As a result, differentiation and mineralization of bone cells, which are suppressed by sclerostin, are upregulated.

5.8.4 Zinc-PolyP

Amorphous Zn-polyP microparticles have been prepared from soluble Na-polyP and $ZnCl_2$ under alkaline conditions and used for integration in collagen-based wound mats (Müller et al. 2020b) and wound gels based on an alginate/oxidized-alginate-gelatin hydrogel matrix (Wang et al. 2020a).

5.8.5 Gadolinium-PolyP

PolyP has also been used as complex with trivalent cations. The combination of polyP with gadolinium (Gd^{3+}) even resulted in a synergistic effect on the stimulation of bone HA formation and ALP activity in SaOS-2, compared with the two components alone (Wang et al. 2016c).

5.8.6 *Amorphous Ca-Phosphate Stabilized with PolyP*

In a further approach, ACP particles have been stabilized with polyP (Müller et al. 2016d, 2020a). The presence of polyP prevents the formation of Ca-phosphate/HA crystals (Müller et al. 2016d). The hybrid particles were prepared by coprecipitation of calcium and phosphate in the presence of 15% polyP. We found that only particles prepared with ≥10% polyP are amorphous and morphogenetically active, while particles obtained in the presence of lower concentrations of polyP (e.g., 5% polyP) are crystalline as demonstrated by XRD analysis (Müller et al. 2016d). The polyP-stabilized ACP turned out to be a suitable matrix that supported not only cell growth and attachment but also exhibited a pronounced osteoblastic and vasculogenic activity. In in vitro experiments, these particles strongly stimulated the mineralization of SaOS-2 cells as well as collagen type 1 and ALP expression (Müller et al. 2016d).

5.9 Biohybrid Formation with Hydrogel-Forming Polymers

Hydrogels are three-dimensional networks formed by cross-linked hydrophilic but water insoluble polymers that are able to absorb large quantities of aqueous fluids (Zhang and Khademhosseini 2017). They can be functionalized by ligands promoting cell attachment or inducing cell growth, differentiation, or specific cellular functions. Therefore, hydrogels are widely used for tissue engineering. Due to their viscoelastic properties, they are also excellently suitable for soft tissue repair (Kocen et al. 2017).

Both natural and synthetic hydrogel-forming polymers are used. These hydrogel-forming polymers comprise both polyanions, polycations, or polymers that carry, at neutral pH, both negative or positive charges. For example, the natural polymers alginate or hyaluronic acid are negatively charged molecules due to their carboxyl groups, while chitosan, which carries multiple amine groups, is a positively charged polymer.

A disadvantage of the hydrogels formed by many hydrogel-forming polymers is a fast degradation rate or a low mechanical strength.

In recent years it has been shown that polyP is ideally suited to be combined with other polyanionic polymers, e.g., *N,O*-carboxymethylchitosan (a negatively charged chitosan derivative); Fig. 5.9. These hybrid materials show the morphogenetic and energy delivering properties of polyP and can be cured by the addition of divalent cations, such as Ca^{2+} or Mg^{2+}, that form metal bridges between the polymers. Thereby, materials with suitable hardness and viscoelastic properties, e.g., for

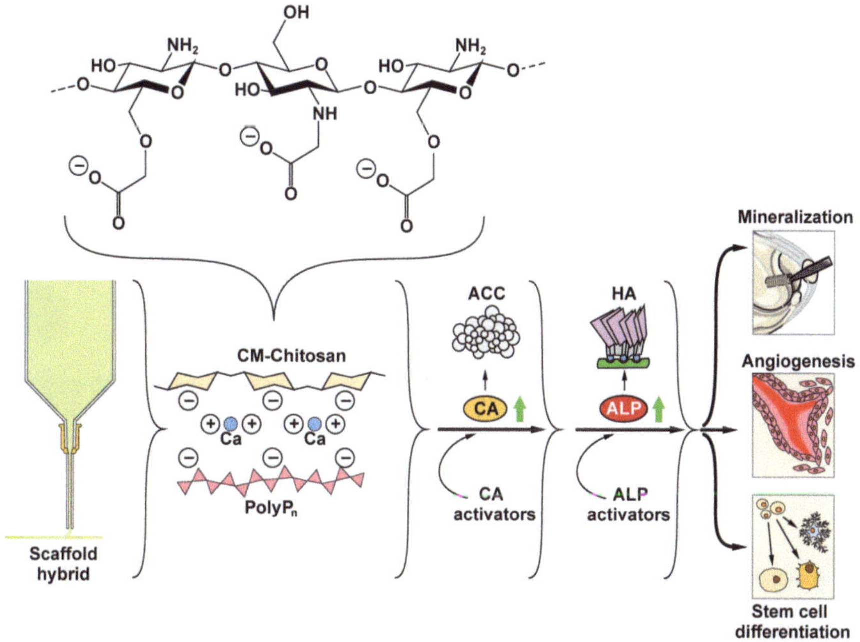

Fig. 5.9 Biohybrid material formation between the two negatively charged polymers polyP and *N*,*O*-carboxymethylchitosan (CM-Chitosan). The printable hybrid material is hardened by printing into a Ca^{2+} containing solution. Both polyanions are cross-linked via Ca^{2+} bridges. The scaffold material promotes bone regeneration/repair by inducing stem cell differentiation, angiogenesis (microvascularization), and mineralization (HA formation). The mineralization process can be additionally increased by supplementation of the material with carbonic anhydrase (CA) or ALP activators

bone and cartilage repair, can be obtained. It is even possible, to use these materials in 3D printing of tailor-made scaffolds, e.g., for manufacturing implants for specific tissue defects.

Using polyP in combination with the likewise polyanionic polymers *N*,*O*-carboxymethylchitosan and alginate, a hydrogel matrix has been obtained that is organized in bundles in the presence of Ca^{2+} ions (Müller et al. 2015d, 2016c); thereby the material hardens due to the formation of cross-links (Ca^{2+} bridges) between the polymers (Fig. 5.10a). In a similar way, a polyP-containing hydrogel matrix has been obtained by cross-linking polyP with hyaluronic acid via Mg^{2+} bridges. This material showed similar viscoelastic properties as cartilage, making it potentially suitable for the therapy of patients suffering from osteoarthrosis (Müller et al. 2016a; Wang et al. 2016a).

Applying a freeze-extraction technique, a macroporous hybrid scaffold has been prepared from polyP, collagen, and chondroitin sulfate (Fig. 5.10b). Addition of calcium ions for Ca^{2+} bridge formation led to the in situ formation of Ca-polyP nanoparticles that decorated the collagen fibers (Müller et al. 2017b). The obtained scaffold exhibited a high porosity, which is required for cell infiltration.

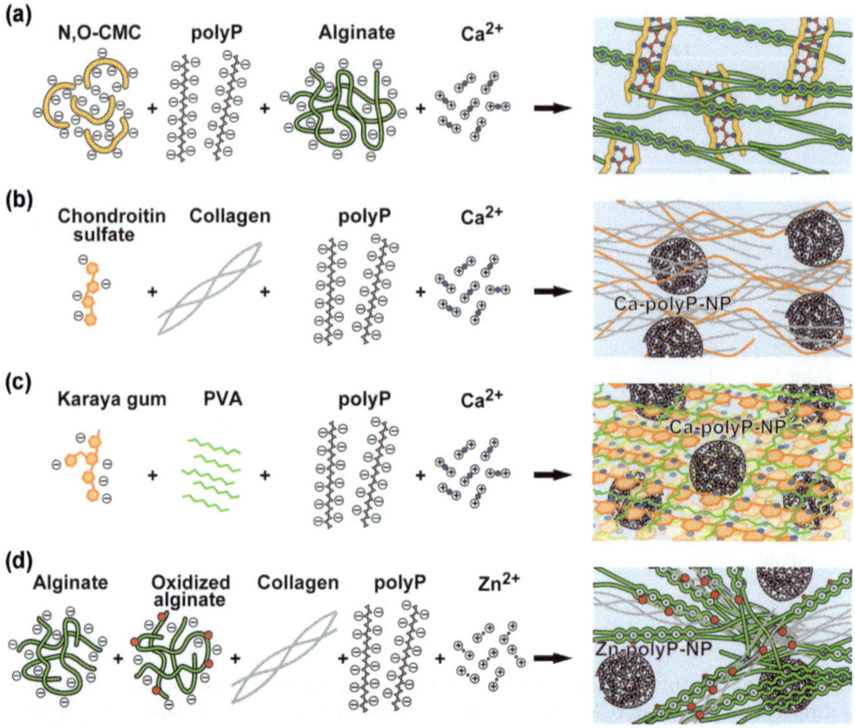

Fig. 5.10 Various hydrogel scaffolds containing polyP or polyP nanoparticles. (**a**) Ca-polyP/*N,O*-carboxymethylchitosan/alginate. The three negatively charged polymers *N,O*-carboxymethylchitosan (*N,O*-CMC), polyP and alginate are organized in bundles in the presence of Ca^{2+} ions. (**b**) Ca-polyP/collagen/chondroitin sulfate. Collagen and chondroitin sulfate are linked via hydrogen bonds. Addition of Ca^{2+} ions leads to formation of Ca^{2+} bridges between polyP and the polysaccharide and to the in situ formation of Ca-polyP nanoparticles. (**c**) Ca-polyP/karaya gum/PVA. The negatively charged karaya gum undergoes Ca^{2+}-mediated ionic gelation (Ca^{2+} bridge formation) with polyP, while PVA forms intermolecular cross-links via hydrogen bonds or crystalline domains between the PVA chains. In addition, Ca-polyP nanoparticles are generated in situ. (**d**) Ca-polyP/alginate/oxidized-alginate/gelatin. Gelatin is bound to a network of alginate and periodate-oxidized alginate, cross-linked via Zn^{2+} ions, by Schiff base reaction of its (hydroxy)-lysine ε-amino groups with the C2 and C3 aldehyde groups of the oxidized alginate. PolyP is integrated either as a Zn-polyP-coacervate via ionic interaction (not shown) or as Zn-polyP nanoparticles formed (shown here). Exposure of the matrices in (**b**)–(**d**) to protein at pH 7 leads to the conversion of the polyP particles in the coacervate phase

Another porous hybrid cryogel material containing in situ generated Ca-polyP nanoparticles has been developed, which is based on karaya gum (a negatively charged polysaccharide hydrocolloid) and poly(vinyl alcohol) (PVA) (Tolba et al. 2018); Fig. 5.10c. The fabrication process of this cryogel included the Ca^{2+}-mediated ionic gelation of the karaya gum and intermolecular cross-linking of PVA by freeze-thawing. This polyP-karaya gum/PVA cryogel showed viscoelastic properties characteristic of cartilage and muscle (Van Loocke et al. 2008; Kocen et al. 2017).

This behavior is most likely due to the different nature of the two kinds of non-covalent bonds formed between the components, the Ca^{2+} bridges between the negatively charged sugar residues of karaya gum and the hydrogen bonds (or crystalline domains) between the PVA chains.

With regard to understanding the functional activity of the material, it was important to demonstrate that the Ca-polyP nanoparticles generated within the cryogel form in a coacervate after exposure to medium/serum (Tolba et al. 2018). This coacervate, the biologically active form of the polymer, was found to attract cells (human MSC) and to support the invasion (and growth) of the cells into the scaffold most likely facilitated by the susceptibility of the ionic and hydrogen bonds between the organic polymers of the cryogel to changes in pH or the activity of enzymes released from the cells (Tolba et al. 2018). The poro-viscoelastic properties of the cryogel were also in a range suitable for vascularization (Strange et al. 2013).

In animal experiments (rat muscle), using microspheres of the polyP-karaya gum/PVA material prepared by an emulsion technology, the implant regions became replaced by initial granulation tissue already after 2 to 4 weeks. After the same implantation period, no tissue formation was observed in controls with microspheres without the material (Tolba et al. 2018).

Based on polyP in combination with hydrogel-forming polymers, it was also possible to develop a suitable bio-ink for 3D bio-printing of living cells. Bone-related SaOS-2 embedded in this bio-ink, consisting of Na-polyP, gelatin, and alginate, complexed with Ca^{2+} ions, were still alive and proliferatively active after printing, due to the energy delivering and morphogenetic properties of the polyP component (Neufurth et al. 2014).

5.10 Application of Biomimetic PolyP-Based Materials in Tissue Regeneration and Repair

PolyP is a nontoxic physiological polymer, which is synthesized throughout the animal kingdom and present in human cells, tissues, and body fluids, even in a nanoparticulate form, like the synthetic polyP nano/microparticles described in this chapter. The unparalleled biomedical properties of this polymer, to be morphogenetically active and deliver metabolic energy, along with its favorable physical-chemical/mechanical properties and the possibility to combine it with other polyanionic, hydrogel-forming polymers, open a number of potential applications in regenerative therapy of bone and cartilage, wound healing, biologizing of surfaces and drug delivery that are described in the following.

PolyP is a safe material. The safety of the polymer has been assessed by the US Food and Drug Administration (FDA) and the European Union; no toxicity has been stated (Food and Drug Administration 1973, 1979). PolyP also exhibits no carcinogenicity, genotoxicity, or reproductive and developmental toxicity (Lanigan 2001; Food and Drug Administration 1975).

5.10.1 Bone

Based on the promising results of the in vitro studies showing that amorphous polyP nano/microparticles exhibit the unique property combination (1) to be morphogenetically active, i.e., to attract (stem) cells and induce cell differentiation via induction of specific transcription factors and growth factors and (2) to deliver metabolic energy required for tissue regeneration, the development of materials for tissue regeneration and repair started. The first focus was on bone regeneration and repair, based on the results that showed that polyP, in addition to its morphogenetic and energy supplying activity, also provides the building material (monomeric orthophosphate) required for HA formation. Using polyP nano/microparticles either alone, after embedding into PLGA [poly(D,L-lactide-*co*-glycolide)] microspheres, or in combination with suitable hydrogel-forming polymers, or integrated into a paste or (bio)printable bio-ink, a number of materials have been developed and tested in ex vivo and in vivo experiments. The aim was to obtain a material that would degrade over time and be replaced by newly formed and functionally active natural bone tissue.

The requirements for an optimal material for bone repair are summarized in Fig. 5.11. Such a material should be able to induce osteogenesis by the recruitment and stimulation of maturation of the precursor cells to the functionally active osteoblasts (osteoinduction), to support bone growth along a surface (osteoconduction) and to allow a stable anchorage of the implant in the surrounding bone tissue (osseointegration) (Albrektsson and Johansson 2001). Only polyP-based materials but neither bio-inert metal (titanium) implants or ceramic implants nor Ca-phosphate/HA-based materials completely fulfil all these requirements.

To study the effect of amorphous Ca-polyP particles on differentiation of bone marrow cells in their physiological microenvironment, femoral explants from mice were used as an ex vivo model (Müller et al. 2018c). The results showed that

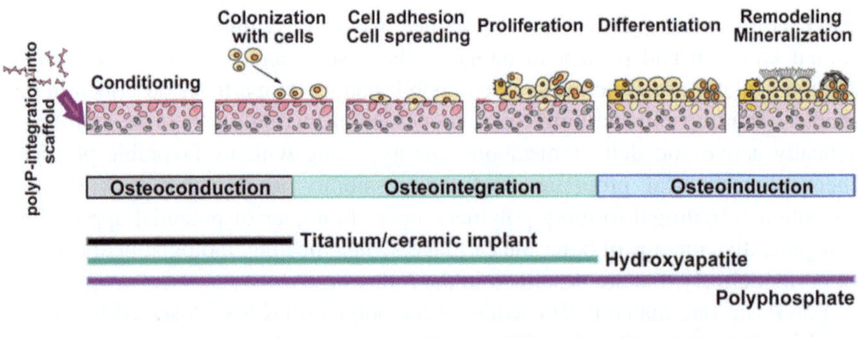

Fig. 5.11 Requirements for an optimal bone implant material. Only polyP-based materials, but not metals/titanium, ceramics, and calcium phosphate/HA, are capable of osteoconduction as well as osseointegration and osteoinduction by promoting cell recruitment, adhesion/spreading, proliferation, and differentiation (morphogenetic effect) as well as remodeling and mineralization (release of metabolic energy and phosphate)

exposure of the explants to the Ca-polyP microparticles resulted in an upregulation of the steady-state expression of the marker genes for the transcription factors SOX9, which promotes differentiation into the chondrogenic lineage, and RUNX2, which induces osteoblast maturation.

Of particular interest for a possible treatment of patients with osteoporosis who have a reduced bone mineral density are results, which showed that the Ca-polyP nano/microparticles also increase the gene expression of OPG without affecting the level of the receptor activator of nuclear factor κB ligand (RANKL). OPG plays a crucial role in the development of the disease. This cytokine acts as an osteoclastogenesis inhibitory factor, which sequesters RANKL and thereby prevents its binding to the receptor RANK. As a result, the maturation of bone-degrading osteoclasts from their precursor cells and osteoclast activity are suppressed (for a review, see Wang et al. 2012; Müller et al. 2013a).

In order to test Ca-polyP nano/microparticles in animal experiments, the particles were embedded, together with polyP-stabilized ACC, into PLGA using an emulsion-based technique (Wang et al. 2016b). The encapsulation prevents the degradation of polyP by the ubiquitously ALP. The 800 μM-sized spheres were implanted into cranial defects of rats; β-TCP was used as control. The results showed that the particles became disintegrated/dissolved during an implantation period of 4 weeks and their space was invaded by cells that were actively synthesizing bone mineral. After 12 weeks, the mechanical properties of the newly formed bone were already close to intact bone (Young's modulus, 1.74 MPa versus 3.05 MPa of normal trabecular bone tissue), while significantly lower values (0.63 mPa) were measured in controls (Wang et al. 2016b). Similar results were obtained with Na-polyP embedded into PLGA microspheres (Wang et al. 2016e).

Next, polyP-based hydrogel materials were developed that are also suitable for use in 3D printing and even in 3D bio-printing (3D printing of living cells). These techniques open up the possibility for future potential manufacture of personalized (patient-specific) regeneratively active implants. By embedding amorphous Ca-polyP particles into a PCL matrix, a 3D-printable bio-ink was developed that enabled the fabrication of tissue-like scaffolds that supported the ingrowth of SaOS-2 cells, most likely by inducing the stromal cell-derived factor-1α (SDF-1α), a cell attracting chemokine (Neufurth et al. 2017). Due to its mechanical properties, PCL has the advantage that it could also be used to manufacture implants for larger bone defects.

In addition to Ca-polyP, animals studies were performed with amorphous Sr-polyP-particles, which, in in vitro experiments, exhibited a higher mineralization inducing activity than the Ca-polyP particles. Again, using the rat critical-size calvarial defect model, it was found that the Sr-polyP particles encapsulated in PLGA microspheres almost completely restored the injured bone area after an implantation period of 12 weeks (Müller et al. 2017e). The healing of the bone defect was even significantly faster than with Ca-polyP particles (Müller et al. 2017e).

Animal experiments also confirmed the promising results from in vitro studies with the polyP-stabilized amorphous ACP particles. For the experiments in rabbits (calvarial bone defect model), the polyP-stabilized ACP (containing 15% polyP) was encapsulated in PLGA microspheres. Crystalline Ca-phosphate (with 5% polyP) and crystalline TCP served as control. The results revealed that the amorphous ACP (15% polyP) caused a pronounced osteoinductive activity already after a six-week healing period, in contrast to the crystalline control particles. In addition, the formation of new bone tissue was paralleled with an increased vascularization and expression of the vascularization marker gene, vascular endothelial growth factor (VEGF) (Müller et al. 2020a). The pronounced anabolic activity of the hybrid material, which combines both osteoinductive and vasculogenic activities, has been attributed to the combined action of both components, ACP and polyP. Ca-phosphate has been reported to act stimulatory on angiogenesis (Malhotra and Habibovic 2016).

The first study on the application of polyP in human bone repair is currently going on (Alkaabi et al. 2021). The aim of this clinical study is to assess the safety and osteoinductivity of amorphous Ca-polyP microparticles as a graft material for alveolar cleft repair.

5.10.2 Cartilage

Cartilage has only a low repair capacity. It is a bradytrophic avascular tissue that contains only a very small number of cells embedded in a large extracellular matrix. Therefore, cartilage, such as articular cartilage, is prone to the development of osteoarthrosis and osteoarthritis. By using amorphous nano/microparticles that contain magnesium instead of calcium, we have succeeded in developing scaffold materials with viscoelastic properties that are very similar to natural cartilage (Müller et al. 2016a; Wang et al. 2016a). In cell culture experiments, the amorphous Mg-polyP nano/microparticles significantly upregulated the growth of human chondrocytes as well as the expression of the marker genes, collagen 3A1 and aggrecan (Müller et al. 2016a; Wang et al. 2016a). The developed regenerative active matrices were based on a combination of the polyanionic polyP and the polyanionic glycosaminoglycan hyaluronic acid, which were cross-linked via Mg^{2+} bridges (Müller et al. 2016a, b, c; Wang et al. 2016a). Because of their beneficial energy providing and morphogenetic properties, as well as their suitable strength and contractibility, these matrices are believed to have great promise for potential use in treatment of osteoarthritis/osteoarthrosis-related cartilage defects.

5.10.3 Wound Healing

The treatment of chronic, nonhealing, or delayed healing wounds, is a medical challenge. Wound healing can be affected under various pathological conditions such as cardiovascular and metabolic disorders like diabetes. Impaired angiogenesis/

neovascularization, which leads to hypoxic cell damage, is a possible cause of chronic wounds in diabetic patients. In addition, there is an increasing incidence of pressure ulcers (decubitus) due to the increasing elderly patients.

Wound healing requires a lot of energy (Bjarnsholt et al. 2008). The lack of energy in patients with diabetes particularly concerns the extracellular space. PolyP has the potential to rise the extracellular ATP level to cover the energy demand of the recovery zone of the injured tissue. The ALP, which produces ATP from polyP through its combined action with ADK, is also present in the wound bed (Müller et al. 2017g). In addition, polyP microparticles can be taken up by cells, resulting in increased intracellular ATP levels (Müller et al. 2017h, 2018a).

In animal experiments, it was found that the topical application of amorphous polyP nano/microparticles significantly accelerates wound healing in normal mice and in diabetic mice with delayed wound healing (Müller et al. 2017d; Wang et al. 2017). The experiments showed that the polyP particles double the reepithelialization rate of experimental wounds already after a healing period of 7 days.

For wound repair gels, polyP can be combined with hydrogel-forming polymers and be applied either as water-soluble Na-polyP or in the form of microparticles, e.g., Ca-polyP of Mg-polyP particles. In the presence of proteins/wound fluid, the polyP is converted into a biologically active coacervate (Müller et al. 2018d), which then induces cell recruitment and differentiation as well as the expression of genes involved in wound healing (Müller et al. 2015b, 2017d). In addition, polyP exhibits antibacterial activity and inhibits matrix metalloproteinases as well as proteases from bacteria, which are involved in delayed wound healing (McCarty et al. 2015).

Based on the successful in vivo data (Müller et al. 2017d) as well as data showing that polyP promotes the growth of keratinocytes in vitro (Simbulan-Rosenthal et al. 2015), a novel collagen-based wound mat containing Zn-polyP microparticles has been developed (Müller et al. 2020b). Zinc ions have been chosen as counterions to polyP because of their beneficial effects on wound healing and tissue repair (Lin et al. 2017), which complement the properties of polyP. In addition, Zn^{2+} is an essential cofactor of ALP (Sorimachi 1987), which is involved in the generation of ATP from the polymer (see above). Zn^{2+} ions also inactivate toxic free radicals and suppress apoptosis (Ogawa et al. 2018).

A disadvantage of conventional collagen hydrogel scaffolds is their low mechanical strength (Ajalloueian et al. 2018). Therefore, we developed a technique whereby the polyP microparticles are integrated into mechanically compressed collagen mats during the compression process (Müller et al. 2020b). The Zn-polyP particles, which were prepared from $ZnCl_2$ and Na-polyP, were found to be amorphous. Integrated into the compressed collagen mats, they are converted into the physiologically active coacervate phase after exposure to peptide-containing body fluids, whereby the polyP becomes biodegradable and unfolds its energy supplying and morphogenetic function.

It should be noted that the epidermis is not vascularized and depends on nutrient delivery from the underlying dermis. The cells of the epidermis derive from stem cells that reside in the basal layer, which is separated from the underlying dermis

(Yokouchi et al. 2016). Therefore, the supply of energy, e.g., by administration of polyP, is of the utmost importance for the regeneration of the epidermis during wound healing. The polyP appears to mimic the physiological role of the platelets during this process. Platelets are rich in polyP, which is secreted after platelet activation (Morrissey et al. 2012; Müller et al. 2015c, 2019a). They are thought to be involved in the four main phases of wound healing, hemostasis, inflammation, proliferation, and maturation/remodeling (Reinke and Sorg 2012). In vitro studies on the effect of the compressed collagen mats with integrated Zn-polyP microparticles showed that these mats clearly stimulated the migration of human skin keratinocytes in these matrices as well as the proliferation and differentiation of the cells (Müller et al. 2020b). The cells embedded in the collagen mats or infiltrating the mats extensively developed microvilli, indicating increased activity of the cells.

In native tissues, the cells are embedded in the adaptable and water-rich hydrogel-like polymer network of the ECM, which provides optimal conditions for cell migration, proliferation, and differentiation. In order to mimic the natural ECM, a biomimetic hydrogel matrix for wound dressings has recently been developed, consisting of a metal (Zn^{2+}) ion cross-linked meshwork of alginate and periodate oxidized alginate, which is bound to gelatin *via* Schiff base formation, and of polyP, which is stably integrated through ionic cross-linking with the Zn^{2+} ions (Wang et al. 2020a); Fig. 5.10d.

The alginate component of this matrix is a copolymer composed of 1,4-linked β-D-mannuronic acid and α-L-guluronic acid. After cross-linking this polysaccharide with divalent cations, a comparatively stiff hydrogel matrix is obtained, which does not allow sufficient cell migration (Schloßmacher et al. 2013). The properties of this matrix can be improved by partial oxidization of the alginate (Reakasame and Boccaccini 2018). The aldehyde groups thereby formed give the oxidized alginate (alginate dialdehyde) the ability to be modified by Schiff's base reaction with amine-containing biomolecules such as proteins (Sarker et al. 2017).

Using sodium metaperiodate as oxidizing agent, a partially oxidized alginate was obtained which was mixed with untreated alginate. After ionic cross-linking with $ZnCl_2$, the hydrogel was exposed to gelatin, resulting in covalent linkage formation by reaction of the ε-amino groups of lysine and hydroxylysine of the protein with the aldehyde groups at C2 and C3 of the alginate sugar units. The integration of polyP provides this alginate/oxidized-alginate-gelatin hydrogel matrix with morphogenetic activity and the ability to act as a donor of metabolic energy to the cells embedded in this matrix. PolyP was integrated either in the form of the physiologically active Zn-polyP-coacervate, formed at neutral pH, or in the form of Zn-polyP nanoparticles, the depot form of the polymer, which is obtained under alkaline conditions (pH 10). After exposure to protein at pH 7, the solid polyP particles are converted into the aqueous coacervate phase (Wang et al. 2020a). In this preparation, Zn^{2+} was used for the ionic cross-linking/interaction of polyP with the alginate/oxidized-alginate-gelatin hydrogel matrix, as this ion is a cofactor of endopeptidases, e.g., matrix metallopeptidase-9, which are essential for cell migration and vascularization during wound healing (Yu et al. 2016).

The experiments revealed that both Zn-polyP-coacervate containing and Zn-polyP-nanoparticle supplemented alginate/oxidized-alginate-gelatin hydrogel matrices significantly promoted the growth of human epidermal keratinocytes as well as the energy-dependent migration and attachment/spreading of the cells, if compared to the matrices lacking the polymer (Wang et al. 2020a). Therefore, these matrices have the potential for an in vivo application in treatment of difficult-to-heal chronic wounds. Studies on patients are going on.

5.10.4 *Microvascularization*

The formation of new blood vessels is an important step in tissue repair to provide the nutrients and oxygen needed for regenerating tissue. Ca-polyP is a potent stimulator of angiogenesis, which starts with the process of microvascularization. It was found that polyP accelerates tube formation of HUVEC cells in the tube formation assay, as a measure for microvascularization (Müller et al. 2018b). Tube formation is an energy-dependent process that requires migration of endothelial cells. In order to show that this process is promoted by polyP via the generation of ATP by the coupled ALP/ADK reaction, inhibitor experiments were carried out (Müller et al. 2018a). The experiments revealed that tube formation induced by the Ca-polyP nano/microparticles is suppressed by levamisole (ALP inhibitor) as well as by Ap_5A (ADK inhibitor). It is also abolished by the addition of apyrase, which causes a depletion of the extracellular ATP pool. These results show that ATP produced from extracellular polyP has an essential role in tube formation. Further experiments revealed that the stimulatory effect of the Ca-polyP nano/microparticles is also impaired by incubating the cells with trifluoperazine (endocytosis blocker) and oxamate (inhibitor of glycolysis). No inhibition of tube formation was observed with oligomycin, an inhibitor of the mitochondrial F_OF_1–ATP synthase, at concentrations that cause a marked reduction of O_2 consumption (Müller et al. 2018a).

From these results, the existence of an autocrine signaling pathway of ATP has been proposed that is involved in the migration of the endothelial cells (Müller et al. 2018a). It is assumed that this process is driven extracellularly by ATP, which is generated from polyP via ALP/ADK, and intracellularly by ATP, which is produced during anaerobic glycolysis.

5.10.5 *Ocular Surface Repair*

Many ocular surface diseases such as diabetic keratopathy or dry eye disease are associated with an impaired regeneration of the corneal epithelium (Villatoro et al. 2017). Due to its energy supplying and morphogenetic functions, polyP might also be useful for topical treatment of corneal epithelial defects. In fact, it was found that polyP promotes the regeneration of the corneal epithelium (Gericke et al. 2019;

Wang et al. 2019). The regeneration starts from the limbal zone at the cornea-conjunctiva border, which harbors the ocular stem cells (Torricelli et al. 2013). These cells divide asymmetrically, forming transient amplifying cells that migrate centripetally and anteriorly and differentiate into the mature corneal epithelial cells (Sagga et al. 2018). Experiments with corneal limbal explants showed an increased outgrowth and migration of corneal epithelial cells when polyP (applied as Na-polyP or polyP particles) was added to the culture medium compared to controls without polyP (Gericke et al. 2019; Wang et al. 2019). In addition, cell cultures experiments revealed that polyP enhances the growth/viability of human corneal epithelial cells (hCECs), as well as the expression of SDF-1 (Gericke et al. 2019) and mucin 1 (Wang et al. 2019).

5.10.6 Surface Biologization

Amorphous Ca-polyP microparticles can also be used for functionalization/biologization of biological inert metals or synthetic fibers. Immobilization of polyP particles to oxidized titanium Ti-6Al-4V via Ca^{2+} bridges was performed using the silane coupling agent (3-aminopropyl)trimethoxysilane (APTMS). The modified surfaces supported the growth of SaOS-2 cells (Müller et al. 2015f). Hernia nets based on synthetic polypropylene fibers were biologized by coating with a polyP/collagen hydrogel by applying a freeze-extraction method (Ackermann et al. 2018). The resulting surface showed deposition of amorphous Ca-polyP microparticles and was found to increase collagen formation and the expression of SDF-1α.

5.10.7 Drug Delivery

PolyP nano/microparticles can also be loaded with drugs for drug delivery. The resulting particles show the combined effects of both polyP and the pharmacologically active ingredients. For example, the incorporation of zoledronic acid, a bisphosphonate drug used for treatment of bone tumors, resulted in particles that showed both the morphogenetic activity of polyP and the growth-inhibitory effect of bisphosphonate (Müller et al. 2018c).

Based on the property of polyP to form under alkaline conditions at a superstoichiometric Ca:P ratio amorphous Ca-polyP nanoparticles and at neutral pH a coacervate, a novel drug encapsulation/delivery technique has been developed (Müller et al. 2020c). Core-shell particles, consisting of a drug-loaded Ca-polyP nanoparticle core surrounded by a drug-loaded Ca-polyP coacervate shell, were stepwise prepared by fabrication of (1) the nanoparticle core from $CaCl_2$ and Na-polyP (2:1 molar ratio) at pH 10, into which dexamethasone (as a phosphate derivative, dexamethasone 21-phosphate) was incorporated, and (2) the coacervate shell by exposure of the Ca-polyP core to soluble Na-polyP and ascorbic acid.

Thereby, the Ca-polyP coacervate shell is formed by the outward migration of Ca^{2+} ions from the Ca-rich Ca-polyP core to the shell. The negatively charged L-ascorbate becomes immobilized, like the glucocorticoid on the polyanionic polyP by Ca^{2+} bridge formation. These core-shell particles caused a strong osteogenic activity in vitro, as a result of the combined effects of the three active ingredients, polyP (morphogenetic and energy delivering function), dexamethasone (osteogenic differentiation), and ascorbic acid (collagen type 1 production). Drug release experiments showed a fast release of the coacervate shell components (ascorbic acid and polyP), while the release of dexamethasone and polyP from the nanoparticle core occurs with a slower rate (Müller et al. 2020c). The latter process involves the conversion of the nanoparticles into the coacervate phase by contact with protein-containing body fluid.

In another approach, amorphous Mg-polyP nanoparticles have been loaded with quercetin (a secondary plant metabolite) or dexamethasone (Neufurth et al. 2021). Human alveolar basal epithelial A549 cells responded to these particles with an upregulation of the expression of the gel-forming mucin MUC5AC, which was significantly higher than the stimulation by the Mg-polyP nanoparticles alone. This result was already found at drug concentrations where both compounds were nearly ineffective. It is assumed that the enhancement of polyP-induced mucin expression by quercetin and the corticosteroid is caused by the scavenging of reactive oxygen species. Both drugs, the polyphenol flavonoid quercetin (Wang et al. 2020b; Kim et al. 2013) and the anti-inflammatory drug dexamethasone (Bas et al. 2012), act as radical scavengers. In addition, the quercetin 3-*O*-glucoside isoquercetin has been described to induce the expression of ALP (Wang et al. 2014b). The generation of ADP/ATP by the combined ALP/ADK action (Müller et al. 2019a) has been proposed to promote the synthesis and organization of the mucin-containing mucus on the respiratory epithelia.

5.10.8 Other Applications

A number of further applications of polyP have been developed, also using additional properties of the polymer. Based on Ca-polyP microparticles, a toothpaste has been prepared, which shows a dual effect on teeth, (1) by sealing cracks or fissures in tooth enamel and dentin via an actively induced remineralization process (Müller et al. 2016b, 2017a; Ackermann et al. 2019) and (2) by killing bacteria, with a pronounced specificity for the cariogenic bacterium *Streptococcus mutans*, in contrast to triclosan, a broad-spectrum antimicrobial agent used in other toothpastes (Müller et al. 2017c). In addition, a polyP-based material for the fabrication of artificial blood vessels has been developed (Neufurth et al. 2015). This material was formed from a hydrogel composed of polyP and two further, negatively charged but inert biopolymers, as well as gelatin, which promotes the binding of endothelial cells via its RGD cell recognition motif. If the Ca^{2+}-free polymer suspension is pressed into a Ca^{2+}-containing hardening solution with a special extruder, small diameter vessels are formed. During this process, a bundling of the initially randomly organized polyanionic polymers in the Ca^{2+}-free suspension under formation of a mechanically stable vascular implant occurs.

With regard to tumor therapy, it has been shown that polyP in human lung A549 (carcinoma) cells induces the expression of the hydrolase that inactivates bleomycin, an anticancer drug that causes strand breaks in DNA after activation (Müller et al. 2021b). Through this mechanism, polyP could help suppress the toxicity of the drug, which can lead to pulmonary fibrosis.

5.11 Future Developments

In the treatment of human diseases, the inorganic biopolymer polyP has proven to be a potentially promising and versatile biomaterial. No other material shows a similar combination of properties as polyP to be morphogenetically active and to provide metabolic energy. In addition, polyP exhibits antibacterial and antiviral properties, even against SARS-CoV-2, the pathogenic agent of COVID-19 disease (see Chap. 7), is a regulator of blood coagulation and fibrinolysis (Morrissey et al. 2012) and probably acts as a chaperone-like molecule (Gray et al. 2014), which could mitigate the toxicity of amyloid proteins in diseases associated with amyloid fibril deposition (Cremers et al. 2016).

With regard to the use of polyP in tissue regeneration/repair, some important areas of possible application have not yet been explored and should be a focus of future research, such as the application of polyP in heart valve tissue engineering. In addition, the benefits of the antiviral/antibacterial activity of polyP should be given more attention, especially with respect to the potential application of polyP in wound healing (wound infection) or treatment of bone defects/fractures (osteomyelitis patients).

PolyP can be used either directly as a biomaterial or component of a biomaterial, or indirectly as a physiologically active substance in the prophylaxis or therapy of bone and cartilage disorders. Possible routes of application comprise the nasal route (via inhalation; see Chap. 7) or the route via the gastrointestinal tract or intravenous injection.

The oral route seems to be the method of choice if polyP is applied in prophylaxis or therapy of osteoporosis. It has been shown that polyP nano/microparticles can be taken up by cells via clathrin-mediated endocytosis (Müller et al. 2017h) or by binding to integrin β1 using the caveolin-1 endocytosis pathway (Tanaka et al. 2015). Even transcellular transport of bacterial polyP nanoparticles across intestinal epithelial cells (Caco-2 cells) has been shown to be possible (Tanaka et al. 2015), but the efficiency of these systems remains to be studied. In particular, the stability of polyP during passage of the gastrointestinal tract has to be evaluated. PolyP will be hydrolyzed under the acidic conditions in the stomach and needs to be stabilized. Methods to prevent the degradation of polyP at acidic pH, e.g., by encapsulation in hydroxypropylmethylcellulose capsules, are available and need to be evaluated.

There is also only limited knowledge about the turnover of polyP both within cells and in extracellular fluids such as in blood. One reason is the lack of sensitive and specific methods for determining and quantifying polyP. Currently, nonradioactive methods are mostly used for polyP analysis, which are less sensitive and often do not allow a definitive proof that the measured material consists exclusively of polyP.

Finally, future directions of research will include the development of polyP-based intelligent scaffold/hydrogel materials that respond to external stimuli from the environment, as well as materials that enable the monitoring of the healing progress in a simple, noninvasive manner. The first steps have already been done with the development of advanced materials (both implants and drug delivery systems) that are capable of in situ formation of polyP nano/microparticles (Müller et al. 2017b; Tolba et al. 2018) or are activated by protein/body fluid contact (Müller et al. 2020b). For example, the wound healing status and the occurrence of wound infections could be monitored by incorporating sensors such as colorimetric sensors into wound gels or mats that can detect relevant parameters for the progress or delay of wound healing, e.g., the pH of the wound bed (optimally pH 4 – 6.5), which becomes more alkaline after bacterial infection, or ALP activity (Deng et al. 2020), which increases during the granulation phase (Alpaslan et al. 1997), or glucose (Huang et al. 2018), which can significantly impair the healing process (diabetic patients) (Müller et al. 2017d).

Such sensors are also needed for intelligent, smart materials for the repair of hard tissue (bone) defects (Ledet et al. 2018). Here, sensors are required that recognize impaired defect healing after implantation, e.g., by recording the stiffness of the material and its reaction to mechanical stress. Sensors, such as piezo-sensors, which allow remote detection of the progress of defect healing would be preferred.

Acknowledgments W.E.G. M. is a holder of an ERC Advanced Investigator Grant (No. 268476). In addition, W.E.G.M. obtained three ERC-PoC grants (Si-Bone-PoC, Grant No. 324564; MorphoVES-PoC, Grant No. 662486; and ArthroDUR, Grant No. 767234). We also acknowledge funding from the European Commission (grants BIO-SCAFFOLDS No. 604036 and BlueGenics No. 311848). Finally, this work was supported by a grant from the Federal Ministry for Economic Affairs and Energy (ZIM—ZF4294001 CS6) and the BiomaTiCS research initiative of the University Medical Center, Mainz.

References

Abramov AY, Fraley C, Diao CT, Winkfein R, Colicos MA, Duchen MR, French RJ, Pavlov E (2007) Targeted polyphosphatase expression alters mitochondrial metabolism and inhibits calcium-dependent cell death. Proc Natl Acad Sci USA 104:18091–18096

Ackermann M, Wang XH, Wang SF, Neufurth M, Schröder HC, Isemer FE, Müller WEG (2018) Collagen-inducing biologization of prosthetic material for hernia repair: polypropylene meshes coated with polyP/collagen. J Biomed Mater Res B Appl Biomater 106:2109–2121

Ackermann M, Tolba E, Neufurth M, Wang S, Schröder HC, Wang XH, Müller WEG (2019) Biomimetic transformation of polyphosphate microparticles during restoration of damaged teeth. Dent Mater 35:244–256

Ajalloueian F, Nikogeorgos N, Ajalloueian A, Fossum M, Lee S, Chronakis IS (2018) Compressed collagen constructs with optimized mechanical properties and cell interactions for tissue engineering applications. Int J Biol Macromol 108:158–166

Albrektsson T, Johansson C (2001) Osteoinduction, osteoconduction and osseointegration. Eur Spine J 10(Suppl 2):S96–101

Alkaabi SA, Natsir Kalla DS, Alsabri GA, Fauzi A, Tajrin A, Müller WEG, Schröder HC, Wang XH, Forouzanfar T, Helder MN, Ruslin M (2021) Polyphosphate (PolyP) for alveolar cleft repair: study protocol for a pilot randomized controlled trial. Trials 22:393

Alpaslan G, Nakajima T, Takano Y (1997) Extracellular alkaline phosphatase activity as a possible marker for wound healing: a preliminary report. J Oral Maxillofac Surg 55:56–62

Angelova PR, Baev AY, Berezhnov AV, Abramov AY (2016) Role of inorganic polyphosphate in mammalian cells: from signal transduction and mitochondrial metabolism to cell death. Biochem Soc Trans 44:40–45

Aoki K, Ogihara N, Tanaka M, Haniu H, Saito N (2020) Carbon nanotube-based biomaterials for orthopaedic applications. J Mater Chem B 8:9227–9238

Bas E, Gupta C, Van De Water TR (2012) A novel organ of corti explant model for the study of cochlear implantation trauma. Anat Rec (Hoboken) 295:1944–1956

Battafarano G, Rossi M, De Martino V, Marampon F, Borro L, Secinaro A, Del Fattore A (2021) Strategies for bone regeneration: from graft to tissue engineering. Int J Mol Sci 22:1128

Bjarnsholt T, Kirketerp-Møller K, Jensen PØ, Madsen KG, Phipps R, Krogfelt K, Høiby N, Givskov M (2008) Why chronic wounds will not heal: a novel hypothesis. Wound Repair Regen 16:2–10

Boes L, Boesebeck H, Ulrich SD, Mont MA, Seyler TM (2009) Collagen-based osteoinductive bone void filler for mis-treatment in osteochondritis dissecans and osteonecrosis. J Bone Joint Surg Br 91-B(Supp II):332

Cescutti G, Matteucci F, Caffau E, François P (2012) Chemical evolution of the Milky Way: the origin of phosphorus. Astron Astrophys 540:A33

Chai JM, Amelia TSM, Mouriya GK, Bhubalan K, Amirul AA, Vigneswari S, Ramakrishna S (2020) Surface-modified highly biocompatible bacterial-poly(3-hydroxybutyrate-*co*-4-hydroxybutyrate): a review on the promising next-generation biomaterial. Polymers 13:51

Chinopoulos C, Vajda S, Csanády L, Mándi M, Mathe K, Adam-Vizi V (2009) A novel kinetic assay of mitochondrial ATP-ADP exchange rate mediated by the ANT. Biophys J 96:2490–2504

Cremers CM, Knoefler D, Gates S, Martin N, Dahl JU, Lempart J, Xie L, Chapman MR, Galvan V, Southworth DR, Jakob U (2016) Polyphosphate: a conserved modifier of amyloidogenic processes. Mol Cell 63:768–780

De Jager HJ, Heyns AM (1998a) Kinetics of acid-catalyzed hydrolysis of a polyphosphate in water. J Phys Chem A 102:2838–2841

De Jager H, Heyns AM (1998b) Study of the hydrolysis of sodium polyphosphate in water using Raman spectroscopy. Appl Spectrosc 52:808–814

Deng HH, Deng Q, Li KL, Zhuang QQ, Zhuang YB, Peng HP, Xia XH, Chen W (2020) Fluorescent gold nanocluster-based sensor for detection of alkaline phosphatase in human osteosarcoma cells. Spectrochim Acta A Mol Biomol Spectrosc 229:117875

Dias Filho FA, Carlos LD, Messadeq Y, Ribeiro SJ (2005) Spectroscopic study and local coordination of polyphosphate colloidal systems. Langmuir 21:1776–1783

Dietzel M (2000) Dissolution of silicates and the stability of polysilicic acid. Geochim Cosmochim Acta 64:3275–3281

Docampo R, Huang G (2016) Acidocalcisomes of eukaryotes. Curr Opin Cell Biol 41:66–72

Docampo R, de Souza W, Miranda K, Rohloff P, Moreno SN (2005) Acidocalcisomes – conserved from bacteria to man. Nat Rev Microbiol 3:251–261

Docampo R, Jimenez V, Lander N, Li ZH, Niyogi S (2013) New insights into roles of acidocalcisomes and contractile vacuole complex in osmoregulation in protists. Int Rev Cell Mol Biol 305:69–113
Fallacara A, Baldini E, Manfredini S, Vertuani S (2018) Hyaluronic acid in the third millennium. Polymers 10:701
Fekry MI, Tipton PA, Gates KS (2011) Kinetic consequences of replacing the internucleotide phosphorus atoms in DNA with arsenic. ACS Chem Biol 6:127–130
Food and Drug Administration (FDA) (1973) GRAS (Generally Recognized as Safe) food ingredients—phosphates. NTIS PB-221-224
Food and Drug Administration (FDA) (1975) Evaluation of the health aspects of food ingredients. National Technical Information Service (NTIS) document number: PB-262-651
Food and Drug Administration (FDA) (1979) Phosphates; proposed affirmation of and deletion from GRAS status as direct and human food ingredients. Federal Register 44(244): 74845–74857, 18 December 1979
Franco DF, De Oliveira Barud HG, Barud HS, Oliveira OB, Meneguin AB, de Oliveira LFC, Silva MAP, Ribeiro SJL, Nalin M (2020) A review on polyphosphate coacervates – structural properties and bioapplications. J Sol-Gel Sci Technol 94:531–543
Frankenthal L (1944) Interaction between sodium triphosphate and salts of polyvalent cations as shown by pH measurements. J Am Chem Soc 66:2124–2126
Frantz C, Stewart KM, Weaver VM (2010) The extracellular matrix at a glance. J Cell Sci 123 (Pt 24):4195–4200
Gericke A, Wang XH, Ackermann M, Wang SF, Neufurth M, Wiens M, Schröder HC, Pfeiffer N, Müller WEG (2019) Utilization of metabolic energy in treatment of ocular surface disorders: polyphosphate as an energy source for corneal epithelial cell proliferation. RSC Adv 9:22531–22539
Glonek T (2021) Did cyclic metaphosphates have a role in the origin of life? Orig Life Evol Biosph 51:1–60
Glowacki J, Mizuno S (2008) Collagen scaffolds for tissue engineering. Biopolymers 89:338–344
Gray MJ, Wholey WY, Wagner NO, Cremers CM, Mueller-Schickert A, Hock NT, Krieger AG, Smith EM, Bender RA, Bardwell JC, Jakob U (2014) Polyphosphate is a primordial chaperone. Mol Cell 53:689–699
Harada K, Itoh H, Kawazoe Y, Miyazaki S, Doi K, Kubo T, Akagawa Y, Shiba T (2013) Polyphosphate-mediated inhibition of tartrate-resistant acid phosphatase and suppression of bone resorption of osteoclasts. PLoS One 8:e78612
Hench LL, Jones JR (2015) Bioactive glasses: frontiers and challenges. Front Bioeng Biotechnol 3: 194
Hothorn M, Neumann H, Lenherr ED, Wehner M, Rybin V, Hassa PO, Uttenweiler A, Reinhardt M, Schmidt A, Seiler J, Ladurner AG, Herrmann C, Scheffzek K, Mayer A (2009) Catalytic core of a membrane-associated eukaryotic polyphosphate polymerase. Science 324: 513–516
Hoylaerts MF, Manes T, Millán JL (1997) Mammalian alkaline phosphatases are allosteric enzymes. J Biol Chem 272:22781–22787
Hu D, Ren Q, Li Z, Zhang L (2020) Chitosan-based biomimetically mineralized composite materials in human hard tissue repair. Molecules 25:4785
Huang Z, Zheng L, Feng F, Chen Y, Wang Z, Lin Z, Lin X, Wenig S (2018) A simple and effective colorimetric assay for glucose based on MnO_2 nanosheets. Sensors (Basel) 18(8). https://doi.org/10.3390/s18082525
Hughes EAB, Robinson TE, Bassett DB, Cox SC, Grover LM (2019) Critical and diverse roles of phosphates in human bone formation. J Mater Chem B 7:7460–7470
Hunter T (2012) Why nature chose phosphate to modify proteins. Philos Trans R Soc Lond B Biol Sci 367:2513–2516
Jeong J, Kim JH, Shim JH, Hwang NS, Heo CY (2019) Bioactive calcium phosphate materials and applications in bone regeneration. Biomater Res 23:4

Ji Y, Zhang S, Qiao M, Jiao R, Li J, Song P, Zhang X, Huang H (2020) Synthesis of structurally defined chondroitin sulfate: paving the way to the structure-activity relationship studies. Carbohydr Polym 248:116796

Kim BH, Choi JS, Yi EH, Lee JK, Won C, Ye SK, Kim MH (2013) Relative antioxidant activities of quercetin and its structurally related substances and their effects on NF-κB/CRE/AP-1 signaling in murine macrophages. Mol Cells 35:410–420

Kirk-Othmer Encyclopedia of Chemical Technology (1991) Phosphoric acid and phosphates, 4th edn, vol 18, pp 669–718

Kocen R, Gasik M, Gantar A, Novak S (2017) Viscoelastic behaviour of hydrogel-based composites for tissue engineering under mechanical load. Biomed Mater 12:025004

Kornberg A (1999) Inorganic polyphosphate: a molecule of many functions. In: Schröder HC, Müller WEG (eds) Inorganic polyphosphates: biochemistry, biology, biotechnology. Prog Molec Subcell Biol 23. Springer, Berlin, pp 1–26

Kornberg A, Rao NN, Ault-Riché D (1999) Inorganic polyphosphate: a molecule of many functions. Ann Rev Biochem 68:89–125

Krasko A, Lorenz B, Batel R, Schröder HC, Müller IM, Müller WEG (2000) Expression of silicatein and collagen genes in the marine sponge *Suberites domuncula* is controlled by silicate and myotrophin. Eur J Biochem 267:4878–4887

Kulaev IS, Vagabov VM, Kulakovskaya TV (2004) The biochemistry of inorganic polyphosphates. Wiley, Chichester, pp 1–277

Kumble KD, Kornberg A (1995) Inorganic polyphosphate in mammalian cells and tissues. J Biol Chem 270:5818–5822

Kumble KD, Kornberg A (1996) Endopolyphosphatases for long chain inorganic polyphosphate in yeast and mammals. J Biol Chem 271:27146–27151

Labberton L, Long AT, Gendler SJ, Snozek CL, Stavrou EX, Nickel KF, Maas C, Blankenberg S, Hernandez JS, Renné T (2018) A flow cytometry-based assay for procoagulant platelet polyphosphate. Cytometry B Clin Cytom 94:369–373

Lander N, Ulrich PN, Docampo R (2013) *Trypanosoma brucei* vacuolar transporter chaperone 4 (TbVtc4) is an acidocalcisome polyphosphate kinase required for *in vivo* infection. J Biol Chem 288:34205–34216

Lanigan RS (2001) Final report on the safety assessment of sodium metaphosphate, sodium trimetaphosphate, and sodium hexametaphosphate. Int J Toxicol 20(Suppl 3):75–89

Lazcano A, Miller SL (1996) The origin and early evolution review of life: prebiotic chemistry, the pre-RNA world, and time. Cell 85:793–798

Ledet EH, Liddle B, Kradinova K, Harper S (2018) Smart implants in orthopedic surgery, improving patient outcomes: a review. Innov Entrep Health 5:41–51

Leyhausen G, Lorenz B, Zhu H, Geurtsen W, Bohnensack R, Müller WEG, Schröder HC (1998) Inorganic polyphosphate in human osteoblast-like cells. J Bone Mineral Res 13:803–812

Lin PH, Sermersheim M, Li H, Lee PHU, Steinberg SM, Ma J (2017) Zinc in wound healing modulation. Nutrients 10:16

Liu W, Zhang X (2015) Receptor activator of nuclear factor-κB ligand (RANKL)/RANK/Osteoprotegerin system in bone and other tissues (review). Mol Med Rep 11:3212–3218

Liu X, Zheng C, Luo X, Wang X, Jiang H (2019) Recent advances of collagen-based biomaterials: multi-hierarchical structure, modification and biomedical applications. Mater Sci Eng C Mater Biol Appl 99:1509–1522

Lorenz B, Schröder HC (2001) Mammalian intestinal alkaline phosphatase acts as highly active exopolyphosphatase. Biochim Biophys Acta 1547:254–261

Lorenz B, Münkner J, Oliveira MP, Kuusksalu A, Leitão JM, Müller WEG, Schröder HC (1997) Changes in metabolism of inorganic polyphosphate in rat tissues and human cells during development and apoptosis. Biochim Biophys Acta 1335:51–60

Lotsari A, Rajasekharan AK, Halvarsson M, Andersson M (2018) Transformation of amorphous calcium phosphate to bone-like apatite. Nat Commun 9:4170

Malhotra A, Habibovic P (2016) Calcium phosphates and angiogenesis: implications and advances for bone regeneration. Trends Biotechnol 34:983–992

Malinauskas T, Jones EY (2014) Extracellular modulators of Wnt signalling. Curr Opin Struct Biol 29:77–84

Mancini I, Defant A (2013) Bioactive poly(arsenic) compounds. Prog Mol Subcell Biol 54:175–195

Mancini I, Guella G, Frostin M, Hnawia E, Laurent D, Debitus C, Pietra F (2006) On the first polyarsenic organic compound from nature: arsenicin A from the new Caledonian marine sponge *Echinochalina bargibanti*. Chemistry 12:8989–8994

Masseron T, García-Hernández DA, Santoveña R, Manchado A, Zamora O, Manteiga M, Dafonte C (2020) Phosphorus-rich stars with unusual abundances are challenging theoretical predictions. Nat Commun 11:3759

McCarty SM, Percival SL, Clegg PD, Cochrane CA (2015) The role of polyphosphates in the sequestration of matrix metalloproteinases. Int Wound J 12:89–99

Millan JL (2006) Alkaline phosphatases - structure, substrate specificity and functional relatedness to other members of a large superfamily of enzymes. Purinergic Signalling 2:335–341

Milo R, Phillips R (2015) Cell biology by the numbers. Garland Science Press, Taylor & Francis Books

Momeni A, Filiaggi MJ (2014) Comprehensive study of the chelation and coacervation of alkaline earth metals in the presence of sodium polyphosphate solution. Langmuir 30:5256–5266

Moore KA, Lemischka IR (2006) Stem cells and their niches. Science 311:1880–1885

Moreno-Sanchez D, Hernandez-Ruiz L, Ruiz FA, Docampo R (2012) Polyphosphate is a novel pro-inflammatory regulator of mast cells and is located in acidocalcisomes. J Biol Chem 287: 28435–28444

Morrissey JH, Choi SH, Smith SA (2012) Polyphosphate: an ancient molecule that links platelets, coagulation, and inflammation. Blood 119:5972–5979

Müller F, Mutch NJ, Schenk WA, Smith SA, Esterl L, Spronk HM, Schmidbauer S, Gahl WA, Morrissey JH, Renné T (2009) Platelet polyphosphates are proinflammatory and procoagulant mediators *in vivo*. Cell 139:1143–1156

Müller WEG, Wang XH, Diehl-Seifert B, Kropf K, Schloßmacher U, Lieberwirth I, Glasser G, Wiens M, Schröder HC (2011) Inorganic polymeric phosphate/polyphosphate as an inducer of alkaline phosphatase and a modulator of intracellular Ca^{2+} level in osteoblasts (SaOS-2 cells) *in vitro*. Acta Biomater 7:2661–2671

Müller WEG, Schröder HC, Schlossmacher U, Grebenjuk VA, Ushijima H, Wang XH (2013a) Induction of carbonic anhydrase in SaOS-2 cells, exposed to bicarbonate and consequences for calcium phosphate crystal formation. Biomaterials 34:8671–8680

Müller WEG, Wang XH, Schröder HC (eds) (2013b) Biomedical inorganic polymers: bioactivity and applications of natural and synthetic polymeric inorganic molecules. Springer, Berlin, vol 54, pp 1–303

Müller WEG, Neufurth M, Huang J, Wang K, Feng Q, Schröder HC, Diehl-Seifert B, Muñoz-Espí R, Wang XH (2015a) Non-enzymatic transformation of amorphous $CaCO_3$ into calcium phosphate mineral after exposure to sodium phosphate *in vitro*: Implications for *in vivo* hydroxyapatite bone formation. ChemBioChem 16:1323–1332

Müller WEG, Tolba E, Dorweiler B, Schröder HC, Diehl-Seifert B, Wang XH (2015b) Electrospun bioactive mats enriched with Ca-polyphosphate/retinol nanospheres as potential wound dressing. Biochem Biophys Rep 3:150–160

Müller WEG, Tolba E, Feng Q, Schröder HC, Markl JS, Kokkinopoulou M, Wang XH (2015c) Amorphous Ca^{2+} polyphosphate nanoparticles regulate ATP level in bone-like SaOS-2 cells. J Cell Sci 128:2202–2207

Müller WEG, Tolba E, Schröder HC, Neufurth M, Wang S, Link T, Al-Nawas B, Wang XH (2015d) A new printable and durable *N,O*-carboxymethyl chitosan-Ca^{2+}-polyphosphate complex with morphogenetic activity. J Mat Chem B 3:1722-1730

Müller WEG, Tolba E, Schröder HC, Wang XH (2015e) Polyphosphate: a morphogenetically active implant material serving as metabolic fuel for bone regeneration. Macromolec Biosci 15: 1182–1197

Müller WEG, Tolba E, Schröder HC, Wang SF, Glasser G, Diehl-Seifert B, Wang XH (2015f) Biologizing titanium alloy implant material with morphogenetically active polyphosphate. RSC Adv 5:75465–75473

Müller WEG, Tolba E, Schröder HC, Wang SF, Glaßer G, Muñoz-Espí R, Link T, Wang XH (2015g) A new polyphosphate calcium material with morphogenetic activity. Mater Lett 148: 163–166

Müller WEG, Ackermann M, Tolba E, Neufurth M, Wang S, Schröder HC, Wang XH (2016a) A bio-imitating approach to fabricate an artificial matrix for cartilage tissue engineering using magnesium-polyphosphate and hyaluronic acid. RSC Adv 6:88559–88570

Müller WEG, Neufurth M, Tolba E, Wang S, Geurtsen W, Feng Q, Schröder HC, Wang XH (2016b) A biomimetic approach to ameliorate dental hypersensitivity by amorphous polyphosphate microparticles. Dent Mater 32:775–783

Müller WEG, Neufurth M, Wang S, Tolba E, Schröder HC and Wang XH (2016c) Morphogenetically active scaffold for osteochondral repair (polyphosphate/alginate/N,O-carboxymethyl chitosan). Eur Cells Mater J [eCM] 31:174-190

Müller WEG, Tolba E, Schröder HC, Muñoz-Espí R, Diehl-Seifert B, Wang XH (2016d) Amorphous polyphosphate-hydroxyapatite: a morphogenetically active substrate for bone-related SaOS-2 cells *in vitro*. Acta Biomater 31:358–367

Müller WEG, Ackermann M, Neufurth M, Tolba E, Wang S, Feng Q, Schröder HC, Wang XH (2017a) A novel biomimetic approach to repair enamel cracks/carious damages and to reseal dentinal tubules by amorphous polyphosphate. Polymers 9:120

Müller WEG, Neufurth M, Ackermann M, Tolba E, Wang S, Feng Q, Schröder HC, Wang XH (2017b) Fabrication of a new physiological macroporous hybrid biomaterial/bioscaffold material based on polyphosphate and collagen by freeze-extraction. J Mat Chem B 5:3823–3835

Müller WEG, Neufurth M, Tolba E, Ackermann M, Korzhev M, Wang S, Feng QL, Schröder HC, Wang XH (2017c) Bifunctional dentifrice: amorphous polyphosphate a regeneratively active sealant with potent anti-*Streptococcus mutans* activity. Dent Mater 33:753–764

Müller WEG, Relkovic D, Ackermann M, Wang S, Neufurth M, Paravic-Radicevic A, Ushijima H, Schröder HC, Wang XH (2017d) Enhancement of wound healing in normal and diabetic mice by topical application of amorphous polyphosphate - superior effect of the host-guest composite material composed of collagen (host) and polyphosphate (guest). Polymers 9:300

Müller WEG, Tolba E, Ackermann M, Neufurth M, Wang S, Feng Q, Schröder HC, Wang XH (2017e) Fabrication of amorphous strontium polyphosphate microparticles that induce mineralization of bone cells *in vitro* and *in vivo*. Acta Biomater 50:89–101

Müller WEG, Wang S, Ackermann M, Neufurth M, Steffen R, Mecja E, Muñoz-Espí R, Feng Q, Schröder HC, Wang XH (2017f) Rebalancing β-amyloid-induced decrease of ATP level by amorphous nano/micro polyphosphate: Suppression of the neurotoxic effect of amyloid β-protein fragment 25-35. Int J Mol Sci 18:2154

Müller WEG, Wang S, Neufurth M, Kokkinopoulou M, Feng Q, Schröder HC, Wang XH (2017g) Polyphosphate as a donor of high-energy phosphate for the synthesis of ADP and ATP. J Cell Sci 130:2747–2756

Müller WEG, Wang S, Wiens M, Neufurth M, Ackermann M, Relkovic D, Kokkinopoulou M, Feng Q, Schröder HC, Wang XH (2017h) Uptake of polyphosphate microparticles *in vitro* (SaOS-2 and HUVEC cells) followed by an increase of the intracellular ATP pool size. PLoS One 12:e0188977

Müller WEG, Ackermann M, Tolba E, Neufurth M, Ivetac I, Kokkinopoulou M, Schröder HC, Wang XH (2018a) Role of ATP during the initiation of microvascularization: acceleration of an autocrine sensing mechanism facilitating chemotaxis by inorganic polyphosphate. Biochem J 475:3255–3273

Müller WEG, Ackermann M, Wang SF, Neufurth M, Muñoz-Espí R, Feng QL, Schröder HC, Wang XH (2018b) Inorganic polyphosphate induces accelerated tube formation of HUVEC endothelial cells. Cell Mol Life Sci 75:21–32

Müller WEG, Neufurth M, Wang S, Ackermann M, Muñoz-Espí R, Feng Q, Lu Q, Schröder HC, Wang XH (2018c) Amorphous, smart, and bioinspired polyphosphate nano/microparticles: a biomaterial for regeneration and repair of osteo-articular impairments *in-situ*. Int J Mol Sci 19: 427

Müller WEG, Wang SF, Tolba E, Neufurth M, Ackermann M, Muñoz-Espí R, Lieberwirth I, Glasser G, Schröder HC, Wang XH (2018d) Transformation of amorphous polyphosphate nanoparticles into coacervate complexes: an approach for the encapsulation of mesenchymal stem cells. Small 14:e1801170

Müller WEG, Schröder HC, Wang XH (2019a) Inorganic polyphosphates as storage for and generator of metabolic energy in the extracellular matrix. Chem Rev 119:12337–12374

Müller WEG, Schröder HC, Wang XH (2019b) The phosphoanhydride bond: one cornerstone of life. The Biochemist 41:22–27

Müller WEG, Ackermann M, Al-Nawas B, Righesso LAR, Muñoz-Espí R, Tolba E, Neufurth M, Schröder HC, Wang XH (2020a) Amplified morphogenetic and bone forming activity of amorphous *versus* crystalline calcium phosphate/polyphosphate. Acta Biomater 118:233–247

Müller WEG, Schepler H, Tolba E, Wang SF, Ackermann M, Muñoz-Espí R, Xiao S, Tan R, She Z, Neufurth M, Schröder HC, Wang H (2020b) A physiologically active interpenetrating collagen network that supports growth and migration of epidermal keratinocytes: zinc-polyP nanoparticles integrated into compressed collagen. J Mater Chem B 8:5892–5902

Müller WEG, Tolba E, Wang SF, Neufurth M, Lieberwirth I, Ackermann M, Schröder HC, Wang XH (2020c) Nanoparticle-directed and ionically forced polyphosphate coacervation: a versatile and reversible core-shell system for drug delivery. Sci Rep 10:17147

Müller WEG, Neufurth M, Lieberwirth I, Muñoz-Espí R, Wang S, Schröder HC, Wang XH (2021a) Triple-target stimuli-responsive anti-COVID-19 face mask with physiological virus-inactivating agents. Biomater Sci 9:6052–6063

Müller WEG, Neufurth M, Wang SF, Schröder HC, Wang XH (2021b) Polyphosphate reverses the toxicity of the quasi-enzyme bleomycin on alveolar endothelial lung cells *in vitro*. Cancers 13: 750

Neubauer VJ, Döbl A, Scheibel T (2021) Silk-based materials for hard tissue engineering. Materials 14:674

Neufurth M, Wang XH, Schröder HC, Feng QL, Diehl-Seifert B, Ziebart T, Steffen R, Wang SF, Müller WEG (2014) Engineering a morphogenetically active hydrogel for bioprinting of bioartificial tissue derived from human osteoblast-like SaOS-2 cells. Biomaterials 35:8810–8819

Neufurth M, Wang XH, Tolba E, Dorweiler B, Schröder HC, Link T, Diehl-Seifert B, Müller WEG (2015) Modular small diameter vascular grafts with bioactive functionalities. PLoS One 10: e0133632

Neufurth M, Wang XH, Wang SF, Steffen R, Ackermann M, Haep ND, Schröder HC, Müller WEG (2017) 3D printing of hybrid (bio)materials for bone tissue engineering: calcium-polyphosphate microparticles encapsulated by polycaprolactone. Acta Biomater 64:377–388

Neufurth M, Wang X, Wang SF, Schröder HC, Müller WEG (2021) Caged dexamethasone/ quercetin nanoparticles, formed of the morphogenetic active inorganic polyphosphate, are strong inducers of *MUC5AC*. Mar Drugs 19:64

Ogawa Y, Kinoshita M, Shimada S, Kawamura T (2018) Zinc and skin disorders. Nutrients 10:199

Omelon SJ, Grynpas MD (2008) Relationships between polyphosphate chemistry, biochemistry and apatite biomineralization. Chem Rev 108:4694–4715

Omelon S, Georgiou J, Henneman ZJ, Wise LM, Sukhu B, Hunt T, Wynnyckyj C, Holmyard D, Ryszard B, Grynpas MD (2009) Control of vertebrate skeletal mineralization by polyphosphates. PLoS One 4:e5634

Pavlov E, Aschar-Sobbi R, Campanella M, Turner RJ, Gómez-García MR, Abramov AY (2010) Inorganic polyphosphate and energy metabolism in mammalian cells. J Biol Chem 285:9420–9428
Poon S, Easterbrook-Smith SB, Rybchyn MS, Carver JA, Wilson MR (2000) Clusterin is an ATP-independent chaperone with very broad substrate specificity that stabilizes stressed proteins in a folding-competent state. Biochemistry 39:15953–15960
Quinn TM, Häuselmann HJ, Shintani N, Hunziker EB (2013) Cell and matrix morphology in articular cartilage from adult human knee and ankle joints suggests depth-associated adaptations to biomechanical and anatomical roles. Osteoarthritis Cartilage 21:1904–1912
Rao NN, Gomez-Garcia MR, Kornberg A (2009) Inorganic polyphosphate: essential for growth and survival. Annu Rev Biochem 78:605–647
Reakasame S, Boccaccini AR (2018) Oxidized alginate-based hydrogels for tissue engineering applications: a review. Biomacromolecules 19:3–21
Reddy MSB, Ponnamma D, Choudhary R, Sadasivuni KK (2021) A comparative review of natural and synthetic biopolymer composite scaffolds. Polymers 13:1105
Reinke JM, Sorg H (2012) Wound repair and regeneration. Eur Surg Res 49:35–43
Reusch R (2014) Biogenesis of ion channels. J Biochem Biophys 1:101
Rezvani Ghomi E, Nourbakhsh N, Akbari Kenari M, Zare M, Ramakrishna S (2021) Collagen-based biomaterials for biomedical applications. J Biomed Mater Res B Appl Biomater 109: 1986–1999
Rice JJ, Martino MM, De Laporte L, Tortelli F, Briquez PS, Hubbell JA (2013) Engineering the regenerative microenvironment with biomaterials. Adv Healthc Mater 2:57–71
Roy S, Guanglei Q, Zuniga-Montanez R, Williams RB, Wuertz S (2021) Recent advances in understanding the ecophysiology of enhanced biological phosphorus removal. Curr Opin Biotechnol 67:166–174
Ruiz FA, Lea CR, Oldfield E, Docampo R (2004) Human platelet dense granules contain polyphosphate and are similar to acidocalcisomes of bacteria and unicellular eukaryotes. J Biol Chem 279:44250–44257
Sagga N, Kuffová L, Vargesson N, Erskine L, Collinson JM (2018) Limbal epithelial stem cell activity and corneal epithelial cell cycle parameters in adult and aging mice. Stem Cell Res 33: 185–198
Sarker B, Singh R, Zehnder T, Forgber T, Alexiou C, Cicha I, Detsch R, Boccaccini AR (2017) Macromolecular interactions in alginate−gelatin hydrogels regulate the behavior of human fibroblasts. J Bioact Compat Polym 32:309–324
Schloßmacher U, Schröder HC, Wang XH, Feng Q, Diehl-Seifert B, Neumann S, Trautwein A, Müller WEG (2013) Alginate/silica composite hydrogel as a potential morphogenetically active scaffold for three-dimensional tissue engineering. RSC Adv 3:11185–11194
Schröder HC, Müller WEG (eds) (1999) Inorganic polyphosphates; biochemistry, biology, biotechnology. Progress in molecular and subcellular biology, vol 23. Springer, Berlin
Schröder HC, Wehland J, Weber K (1985) Purification of brain tubulin-tyrosine ligase by biochemical and immunological methods. J Cell Biol 100:276–281
Schröder HC, Kurz L, Müller WEG, Lorenz B (2000) Polyphosphate in bone. Biochemistry (Moscow) 65:296–303
Schröder HC, Wang XH, Manfrin A, Yu SH, Grebenjuk VA, Korzhev M, Wiens M, Schloßmacher U, Müller WEG (2012) Silicatein: acquisition of structure-guiding and structure-forming properties during maturation from the pro-silicatein to the silicatein form. J Biol Chem 287:22196–22205
Schröder HC, Grebenjuk VA, Wang XH, Müller WEG (2016) Hierarchical architecture of sponge spicules: Biocatalytic and structure-directing activity of silicatein proteins as model for bioinspired applications. Bioinspir Biomim 11:041002
Schwiebert EM, Zsembery A (2003) Extracellular ATP as a signaling molecule for epithelial cells. Biochim Biophys Acta 1615:7–32

Seidlmayer LK, Gomez-Garcia MR, Blatter LA, Pavlov E, Dedkova EN (2012) Inorganic polyphosphate is a potent activator of the mitochondrial permeability transition pore in cardiac myocytes. J Gen Physiol 139:321–231

Sheldon KL, Gurnev PA, Bezrukov SM, Sackett DL (2015) Tubulin tail sequences and post-translational modifications regulate closure of mitochondrial voltage-dependent anion channel (VDAC). J Biol Chem 290:26784–26789

Silva MA, Franco DF, de Oliveira LF (2008) New insight on the structural trends of polyphosphate coacervation processes. J Phys Chem A 112:5385–5389

Simbulan-Rosenthal CM, Gaur A, Sanabria VA, Dussan LJ, Saxena R, Schmidt J, Kitani T, Chen YS, Rahim S, Uren A, Crooke E, Rosenthal DS (2015) Inorganic polyphosphates are important for cell survival and motility of human skin keratinocytes. Exp Dermatol 24:636–639

Smith SA, Mutch NJ, Baskar D, Rohloff P, Docampo R, Morrissey JH (2006) Polyphosphate modulates blood coagulation and fibrinolysis. Proc Natl Acad Sci USA 103:903–908

Sorimachi K (1987) Activation of alkaline phosphatase with Mg^{2+} and Zn^{2+} in rat hepatoma cells. Accumulation of apoenzyme. J Biol Chem 262:1535–1541

Strange DGT, Fletcher TL, Tonsomboon K, Brawn H, Zhao X, Oyen ML (2013) Separating poroviscoelastic deformation mechanisms in hydrogels. Appl Phys Lett 102:031913

Tagliabracci VS, Engel JL, Wen J, Wiley SE, Worby CA, Kinch LN, Xiao J, Grishin NV, Dixon JE (2012) Secreted kinase phosphorylates extracellular proteins that regulate biomineralization. Science 336:1150–1153

Tagliabracci VS, Pinna LA, Dixon JE (2013) Secreted protein kinases. Trends Biochem Sci 38: 121–130

Tagliabracci VS, Wiley SE, Guo X, Kinch LN, Durrant E, Wen J, Xiao J, Cui J, Nguyen KB, Engel JL, Coon JJ, Grishin N, Pinna LA, Pagliarini DJ, Dixon JE (2015) A single kinase generates the majority of the secreted phosphoproteome. Cell 161:1619–1632

Tammenkoski M, Koivula K, Cusanelli E, Zollo M, Steegborn C, Baykov AA, Lahti R (2008) Human metastasis regulator protein H-prune is a short-chain exopolyphosphatase. Biochemistry 47:9707–9713

Tanaka K, Fujiya M, Konishi H, Ueno N, Kashima S, Sasajima J, Moriichi K, Ikuta K, Tanabe H, Kohgo Y (2015) Probiotic-derived polyphosphate improves the intestinal barrier function through the caveolin-dependent endocytic pathway. Biochem Biophys Res Commun 467: 541–548

Theocharis AD, Manou D, Karamanos NK (2019) The extracellular matrix as a multitasking player in disease. FEBS J 286:2830–2869

Thilo E, Wicker W (1957) Zur Chemie der kondensierten Phosphate und Arsenate. XVIII. Anionenhydrolyse der kondensierten Phosphate in verdünnter wässriger Lösung. Z Anorg Allg Chem 291:164–185

Tolba E, Müller WEG, El-Hady BMA, Neufurth M, Wurm F, Wang S, Schröder HC, Wang XH (2016) High biocompatibility and improved osteogenic potential of amorphous calcium carbonate/vaterite. J Mat Chem B 4:376–386

Tolba E, Wang XH, Ackermann M, Neufurth M, Muñoz-Espí R, Schröder HC, Müller WEG (2018) *In-situ* polyphosphate nanoparticle formation in hybrid poly(vinyl alcohol)/karaya gum-hydrogels: a porous scaffold inducing infiltration of mesenchymal stem cells. Adv Sci 2018:1801452

Torricelli AA, Singh V, Santhiago MR, Wilson SE (2013) The corneal epithelial basement membrane: structure, function, and disease. Invest Ophthalmol Vis Sci 54:6390–6400

Ulrich PN, Lander N, Kurup SP, Reiss L, Brewer J, Soares Medeiros LC, Miranda K, Docampo R (2014) The acidocalcisome vacuolar transporter chaperone 4 catalyzes the synthesis of polyphosphate in insect-stages of *Trypanosoma brucei* and *T. cruzi*. J Eukaryot Microbiol 61: 155–165

Upadhyay RK (2017) Role of calcium bio-minerals in regenerative medicine and tissue engineering. J Stem Cell Res Ther 2:00081

Usui Y, Uematsu T, Uchihashi T, Takahashi M, Takahashi M, Ishizuka M, Doto R, Tanaka H, Komazaki Y, Osawa M, Yamada K, Yamaoka M, Furusawa K (2010) Inorganic polyphosphate induces osteoblastic differentiation. J Dent Res 89:504–509

Van Loocke M, Lyons CG, Simms CK (2008) Viscoelastic properties of passive skeletal muscle in compression: stress-relaxation behaviour and constitutive modelling. J Biomech 41:1555–1566

Van Wazer JR (ed) (1958) Phosphorus and its compounds: chemistry, vol 1. Interscience Publishers, New York

Van Wazer JR, Campanella DA (1950) Structure and properties of the condensed phosphates. IV. Complex ion formation in polyphosphate solutions. J Am Chem Soc 72:655–663

Verhoef JJ, Barendrecht AD, Nickel KF, Dijkxhoorn K, Kenne E, Labberton L, McCarty OJ, Schiffelers R, Heijnen HF, Hendrickx AP, Schellekens H, Fens MH, de Maat S, Renné T, Maas C (2017) Polyphosphate nanoparticles on the platelet surface trigger contact system activation. Blood 129:1707–1717

Villatoro AJ, Fernández V, Claros S, Alcoholado C, Cifuentes M, Merayo-Lloves J, Andrades JA, Becerra J (2017) Regenerative therapies in dry eye disease: from growth factors to cell therapy. Int J Mol Sci 18(11)

Von Euw S, Wang Y, Laurent G, Drouet C, Babonneau F, Nassif N, Azaïs T (2019) Bone mineral: new insights into its chemical composition. Sci Rep 9:8456

Walsh CT (2020) The chemical biology of phosphorus. Royal Society of Chemistry; ISBN: 978-1-83916-202-2, pp 1–546

Wang L, Fraley CD, Faridi J, Kornberg A, Roth RA (2003) Inorganic polyphosphate stimulates mammalian TOR, a kinase involved in the proliferation of mammary cancer cells. Proc Natl Acad Sci USA 100:11249–11254

Wang XH, Schröder HC, Wiens M, Ushijima H, Müller WEG (2012) Bio-silica and bio-polyphosphate: applications in biomedicine (bone formation). Curr Opin Biotechnol 23: 570–578

Wang XH, Schröder HC, Diehl-Seifert B, Kropf K, Schlossmacher U, Wiens M, Müller WEG (2013a) Dual effect of inorganic polymeric phosphate/polyphosphate on osteoblasts and osteoclasts *in vitro*. J Tissue Engin Regen Med 7:767–776

Wang XH, Schröder HC, Feng QL, Draenert F, Müller WEG (2013b) The deep-sea natural products, biogenic polyphosphate (Bio-PolyP) and biogenic silica (Bio-Silica), as biomimetic scaffolds for bone tissue engineering: fabrication of a morphogenetically-active polymer. Mar Drugs 11:718–746

Wang XH, Schröder HC, Schloßmacher U, Müller WG (2013c) Inorganic polyphosphates: biologically active biopolymers for biomedical applications. In: Biomedical inorganic polymers: bioactivity and applications of natural and synthetic polymeric inorganic molecules (Müller WEG, Wang XH, Schröder HC, eds), Springer-Press, Berlin. Prog Mol Subcell Biol 54:261–294

Wang SF, Wang XH, Draenert FG, Albert O, Schröder HC, Mailänder V, Mitov G, Müller WEG (2014a) Bioactive and biodegradable silica biomaterial for bone regeneration. Bone 67:292–304

Wang XH, Schröder HC, Feng Q, Diehl-Seifert B, Grebenjuk VA, Müller WEG (2014b) Isoquercitrin and polyphosphate co-enhance mineralization of human osteoblast-like SaOS-2 cells via separate activation of two RUNX2 cofactors AFT6 and Ets1. Biochem Pharmacol 89: 413–421

Wang XH, Schröder HC, Grebenjuk V, Diehl-Seifert B, Mailänder V, Steffen R, Schloßmacher U, Müller WEG (2014c) The marine sponge-derived inorganic polymers, biosilica and polyphosphate, as morphogenetically active matrices/scaffolds for the differentiation of human multipotent stromal cells: potential application in 3D printing and distraction osteogenesis. Mar Drugs 12:1131–1147

Wang XH, Schröder HC, Müller WEG (2014d) Enzymatically synthesized inorganic polymers as morphogenetically active bone scaffolds: application in regenerative medicine. Int Rev Cell Mol Biol 313:27–77

Wang XH, Schröder HC, Müller WEG (2014e) Enzyme-based biosilica and biocalcite: biomaterials for the future in regenerative medicine. Trends Biotechnol 32:441–447
Wang XH, Schröder HC, Schlossmacher U, Neufurth M, Feng QL, Diehl-Seifert B, Müller WEG (2014f) Modulation of the initial mineralization process of SaOS-2 cells by carbonic anhydrase activators and polyphosphate. Calcif Tissue Int 94:495–509
Wang XH, Ackermann M, Tolba E, Neufurth M, Wurm F, Feng Q, Wang S, Schröder HC, Müller WEG (2016a) Artificial cartilage bio-matrix formed of hyaluronic acid and Mg^{2+}-polyphosphate. Eur Cells Mater 32:271–283
Wang XH, Ackermann M, Wang S, Tolba E, Neufurth M, Feng Q, Schröder HC, Müller WEG (2016b) Amorphous polyphosphate/amorphous calcium carbonate implant material with enhanced bone healing efficacy in a critical-size-defect in rats. Biomed Mater 11:035005
Wang XH, Huang J, Wang K, Neufurth M, Schröder HC, Wang SF, Müller WEG (2016c) The morphogenetically active polymer, inorganic polyphosphate complexed with $GdCl_3$, as an inducer of hydroxyapatite formation *in vitro*. Biochem Pharmacol 102:97–106
Wang XH, Schröder HC, Müller WEG (2016d) Polyphosphate as a metabolic fuel in Metazoa: a foundational breakthrough invention for biomedical applications. Biotechnol J 11:11–30
Wang XH, Wang S, He F, Tolba E, Schröder HC, Diehl-Seifert B, Müller WEG (2016e) Polyphosphate as a bioactive and biodegradable implant material: induction of bone regeneration in rats. Adv Eng Mat 18:1406–1417
Wang XH, Ackermann M, Neufurth M, Wang SF, Li Q, Feng QL, Schröder HC, Müller WEG (2017) Restoration of impaired metabolic energy balance (ATP pool) and tube formation potential of endothelial cells under "high glucose", diabetic conditions by the bioinorganic polymer polyphosphate. Polymers 9:575
Wang XH, Schröder HC, Müller WEG (2018) Amorphous polyphosphate, a smart bioinspired nano-/bio-material for bone and cartilage regeneration: towards a new paradigm in tissue engineering. J Mat Chem B 6:2385–2412
Wang XH, Gericke A, Ackermann M, Wang SF, Neufurth M, Schröder HC, Pfeiffer N, Müller WEG (2019) Polyphosphate, the physiological metabolic fuel for corneal cells: a potential biomaterial for ocular surface repair. Biomater Sci 7:5506–5515
Wang SF, Wang XH, Neufurth M, Tolba E, Schepler H, Xiao S, Schröder HC, Müller WEG (2020a) Biomimetic alginate/gelatin cross-linked hydrogels supplemented with polyphosphate for wound healing applications. Molecules 25:5210
Wang Y, Tao B, Wan Y, Sun Y, Wang L, Sun J, Li C (2020b) Drug delivery based pharmacological enhancement and current insights of quercetin with therapeutic potential against oral diseases. Biomed Pharmacother 128:110372
Weiner S, Mahamid J, Politi Y, Ma Y, Addadi L (2009) Overview of the amorphous precursor phase strategy in biomineralization. Front Mater Sci China 3:104
Westheimer FH (1987) Why nature chose phosphates. Science 235:1173–1178
Wijeyewickrema LC, Lameignere E, Hor L, Duncan RC, Shiba T, Travers RJ, Kapopara PR, Lei V, Smith SA, Kim H, Morrissey JH, Pike RN, Conway EM (2016) Polyphosphate is a novel cofactor for regulation of complement by a serpin, C1 inhibitor. Blood 128:1766–1776
Wilson MR, Easterbrook-Smith SB (2000) Clusterin is a secreted mammalian chaperone. Trends Biochem Sci 25:95–98
Wyatt AR, Yerbury JJ, Berghofer P, Greguric I, Katsifis A, Dobson CM, Wilson MR (2011) Clusterin facilitates *in vivo* clearance of extracellular misfolded proteins. Cell Mol Life Sci 68: 3919–3931
Yewdall NA, André AAM, Lu T, Spruijt E (2021) Coacervates as models of membrane-less organelles. Curr Opin Colloid Interface Sci 52:101416

Yokouchi M, Atsugi T, Logtestijn MV, Tanaka RJ, Kajimura M, Suematsu M, Furuse M, Amagai M, Kubo A (2016) Epidermal cell turnover across tight junctions based on Kelvin's tetrakaidecahedron cell shape. Elife 5:e19593
Yu S, Sathi GA, Yamamoto O (2016) *In vivo* evaluation of wound healing property of zinc smectite using a rat model. J Ceramic Soc Jpn 124:1199–1204
Zarrintaj P, Manouchehri S, Ahmadi Z, Saeb MR, Urbanska AM, Kaplan DL, Mozafari M (2018) Agarose-based biomaterials for tissue engineering. Carbohydr Polym 187:66–84
Zhang YS, Khademhosseini A (2017) Advances in engineering hydrogels. Science 356: eaaf3627

Chapter 6
Effects of Polyphosphate on Leukocyte Function

Patrick M. Suess

Abstract Leukocytes are immune cells derived from hematopoietic stem cells of the bone marrow which play essential roles in inflammatory and immune responses. In contrast to anucleate platelets and erythrocytes, leukocytes are differentiated from other blood cells by the presence of a nucleus, and consist of monocytes, neutrophils, lymphocytes, basophils, and eosinophils. Factors released from platelets mediate immune responses in part by recruitment and regulation of leukocyte activity. Platelet dense granules contain the highly anionic polymer polyphosphate (polyP) with monomer chain lengths of approximately 60–100 phosphates long, which are released into the microenvironment upon platelet activation. Recent studies suggest that polyP released from platelets plays roles in leukocyte migration, recruitment, accumulation, differentiation, and activation. Furthermore, bacterial-derived polyphosphate, generally consisting of phosphate monomer lengths in the hundreds to thousands, appear to play a role in pathogenic evasion of the host immune response. This review will discuss the effects of host and pathogenic-derived polyphosphate on leukocyte function.

Keywords Polyphosphate · leukocytes · platelets

6.1 Introduction

Inorganic polyphosphate (polyP) is a highly anionic polymer found throughout nature (Brown and Kornberg 2004). PolyP consists of linear chains of orthophosphates bound by high-energy phosphoanhydride bonds, varying in length from a few phosphate monomers to over a thousand depending on the organism (Rao et al. 2009). While remarkably simple in composition and structure, polyP has a myriad of roles in cell biology. The role of polyP in prokaryotic and unicellular eukaryotic organisms has been well studied. Prokaryotes and the eukaryotic social amoeba,

P. M. Suess (✉)
Department of Biological Chemistry, University of Michigan Medical School, Ann Arbor, MI, USA
e-mail: psuess@med.umich.edu

W. E. G. Müller et al. (eds.), *Inorganic Polyphosphates*, Progress in Molecular and Subcellular Biology 61, https://doi.org/10.1007/978-3-031-01237-2_6

Dictyostelium discoideum, metabolize polyP by transferring the terminal phosphate of ATP or GTP to the growing end of a polyP chain using the enzyme polyphosphate kinase (PPK) in a reversible reaction (Rao et al. 2009). Microbial polyP plays roles in quorum sensing, proliferation, survival, stress endurance, motility, virulence, metal chelation, and development (Rao and Kornberg 1996; Rashid et al. 2000a, b; Nikel et al. 2013; Livermore et al. 2016; Singh et al. 2016; Suess and Gomer 2016; Suess et al. 2017; Kulakovskaya 2018).

The roles of polyP in mammalian cell biology, relative to microbial cells, are less understood. The enzyme(s) responsible for polyP synthesis in mammals has not been identified, however, recent work from Baev et al. found that the mammalian F_0F_1-ATP synthase is capable of synthesizing and hydrolyzing polyphosphate in the mitochondria, marking the first instance of a mammalian enzyme with this activity (Baev et al. 2020). The lack of known genetic regulators of polyP metabolism has prevented genetic manipulation from impacting polyP production and metabolism, hindering studies of polyP in mammalian cells. However, in 2004 Ruiz and colleagues identified polyP within the dense granules of human platelets, where it is released into the microenvironment upon platelet activation (Ruiz et al. 2004). These findings catalyzed numerous works on the role of platelet-derived polyP, which is now known to play roles in coagulation, migration, inflammation, wound healing, and differentiation (Smith et al. 2006; Usui et al. 2010; Hassanian et al. 2015; Müller et al. 2017, 2018; Suess et al. 2020). While the enzyme responsible for polyP synthesis in platelets remains unknown, in 2013 Ghosh et al. reported that mice lacking the inositol hexakisphosphate kinase 1 (Ip6k1) contain abnormally low levels of platelet-polyP (Ghosh et al. 2013), suggesting a role for this enzyme.

Platelet-derived factors play an important role in regulating immune cell function and activity, however, the role of platelet-polyP in this capacity has only recently been investigated (Morrell et al. 2014; Ed Rainger et al. 2015; Kral et al. 2016). This review will discuss the known effects of polyP on leukocyte function, detailing the molecular mechanisms by which polyP can modulate immune responses.

6.2 Sources of PolyP Exposure

Platelets, while most often associated with their role in thrombosis and hemostasis, play an important role in modulating innate immune responses (Jenne and Kubes 2015; Rossaint et al. 2018). One such form of regulation is through the release of granule contents such as growth factors, chemokines, and cytokines upon platelet activation. These factors impact leukocyte recruitment, accumulation, and activity (Ed Rainger et al. 2015). PolyP ranging in length from roughly 60 to 100 phosphate monomers is contained within the dense granules of platelets, and upon activation is released into the extracellular environment and modulates coagulation and fibrinolysis (Ruiz et al. 2004; Smith et al. 2006). In addition to roles in blood coagulation, platelet-polyP is an emerging cell signaling molecule eliciting pro-inflammatory and pro-coagulant effects on endothelial cells, and pro-wound healing effects on fibroblasts (Dinarvand et al. 2014; Hassanian et al. 2015; Biswas et al. 2018; Carney et al. 2020; Suess et al. 2020). These findings have prompted interest in potential roles for platelet-derived polyP on leukocyte activity and function.

Other blood cells also contain granular polyP reserves. Granulocytic mast cells and basophils contain granular polyP of approximately the same size as platelet-polyP, which can be released upon degranulation (Moreno-Sanchez et al. 2012). Like platelets, factors released by these cells have modulatory effects on immune responses (Cardamone et al. 2016). Peripheral blood mononuclear cells (PBMCs) consist of lymphocytes, monocytes, and dendritic cells, and have been shown to contain intracellular polyP (Lorenz et al. 1997), however, the subcellular localization, potential release, and impact of polyP in these cells has not been investigated.

Microbes also accumulate intracellular polyP, generally of longer lengths relative to mammalian-derived polyP, reaching up to over a thousand phosphate monomers in some cases (Rao et al. 2009). Certain pathogenic organisms such as *Mycobacterium smegmatis* and *Porphyromonas gingivalis* have been shown to produce and accumulate extracellular polyP (Neilands and Kinnby 2020; Rijal et al. 2020). In other cases, bacteria such as *Neisseria meningitidis* contain polyP as a capsule-like coating attached to the cell surface (Tinsley et al. 1993).

These findings suggest that host immune cells likely encounter both host-derived extracellular short-chain polyP from platelets, mast cells, and basophils, and long-chain polyP produced by microbes.

6.3 PolyP Effects on Neutrophils

Neutrophils are a type of polymorphonuclear leukocyte that play an essential role in inflammation, innate immunity, and thrombosis, and are among the first leukocytes to be recruited to a site of inflammation (Kolaczkowska and Kubes 2013). Platelet and neutrophil interactions are critical to early responses to injury or infection (Iba and Levy 2018). Platelet-polyP has emerged as a mediator of these interactions and subsequent neutrophil activity.

Bacterial infections induce increases in neutrophil accumulation that can be facilitated by neutrophil-platelet aggregates (NPAs) (Finsterbusch et al. 2018). Platelets deficient in Ip6k1 contain abnormally low granular polyP content, providing a genetic background in which decreased levels of platelet-polyP can be assessed (Ghosh et al. 2013). Ip6k1-deficient mice show decreased neutrophil accumulation and NPA formation in a bacterial pneumonia model due to a lack of Ip6k1 in platelets rather than in neutrophils (Hou et al. 2018). The authors investigated the possible role of platelet-polyP in this defect using an ex vivo system in which purified platelets and neutrophils from WT and Ip6k1-deficient backgrounds were used to assess aggregation in response to lipopolysaccharide (LPS). LPS-induced NPAs were abolished when platelets lacking Ip6k1 were incubated with either WT or Ip6k1-deficient neutrophils (Hou et al. 2018). LPS stimulation increased polyP secretion in neutrophil-platelet co-cultures, and this activity was suppressed when

Ip6k1 was disrupted in platelets, but not in neutrophils (Hou et al. 2018). Defective NPA formation after LPS stimulation was rescued upon addition of exogenous polyP (Hou et al. 2018). Consistent with these findings, injection of polyP into Ip6k1-deficient mice rescued LPS-induced NPA formation and neutrophil accumulation (Hou et al. 2018). Interestingly, polyP treatment enhanced neutrophil accumulation even in the absence of NPA formation, suggesting additional mechanisms independent of NPA formation were at play (Hou et al. 2018). These findings suggest that platelet-polyP mediates neutrophil recruitment and accumulation upon bacterial infection.

PolyP also influences the interaction of neutrophils with the endothelium, as intrascrotal injection of polyP enhanced neutrophil recruitment and neutrophil–endothelial cell interactions in the mouse cremaster microvasculature (Du et al. 2019). The authors uncovered the molecular mechanism by which polyP alters neutrophil recruitment, finding that immunoneutralization of P-selectin or PSGL-1 drastically reduced polyP-induced neutrophil rolling, adhesion, and emigration, while Mac-1 and LFA-1 immunoneutralization decreased neutrophil adhesion and emigration, but not rolling (Du et al. 2019). In line with these studies, polyP was found to act as a chemoattractant for human neutrophils in an in vitro model of neutrophil migration (Suess et al. 2019). These findings further enforce the notion that polyP is a powerful inducer of neutrophil recruitment and accumulation.

Neutrophils are phagocytic, capable of ingesting and killing invading pathogens, however, neutrophils have also developed other means by which to combat pathogens, such as release of Neutrophil Extracellular Traps (NETs), which are composed of DNA, histones, and granular contents (Kolaczkowska and Kubes 2013). NET formation aids in preventing bacterial dispersal by binding and killing bacteria as well as degrading virulence factors (Iba and Levy 2018). TLR4-mediated activated platelets bind to neutrophils and lead to neutrophil activation and NET formation (Clark et al. 2007). The role of platelet-polyP in this process has only recently been investigated. In 2017, Chrysanthopoulou et al. observed that synthetic polyP induced NET formation in human neutrophils in a dose-dependent manner (Chrysanthopoulou et al. 2017). Neutrophils incubated with supernatants from activated platelets produced NETs, and the NET inducing activity was abrogated when supernatants were treated with alkaline phosphatase, an enzyme that can degrade polyP, suggesting that polyP is the active inducer (Chrysanthopoulou et al. 2017). In line with this observation, confocal microscopy revealed that platelet-derived polyP localized in close proximity to neutrophil and NET remnants (Chrysanthopoulou et al. 2017). Autophagy is an important step in NET release. PolyP treatment of neutrophils inhibited mTOR signaling, thereby promoting autophagy and NETosis (Chrysanthopoulou et al. 2017). In accordance with these findings, in 2019 Madhi et al. observed that activated Ip6k1-deficient platelets, which contain abnormally low polyP levels, are attenuated in their ability to induce NET formation relative to activated WT platelets (Madhi et al. 2019). This defect of Ip6k1-deficient platelets was rescued upon exogenous addition of polyP (Madhi et al. 2019). These findings suggest that platelet-polyP mediates thromboinflammation in part by regulating NET formation.

The role of platelets in modulating neutrophil activity has long been documented, however, the contribution of platelet-derived polyP to these functions has only begun to be investigated. These reports suggest that platelet-polyP plays a profound role in this mediation.

6.4 PolyP Effect on Monocytes and Macrophages

Monocytes are a class of white blood cell that originates in the bone marrow and shares a common myeloid progenitor with neutrophils (Gordon and Taylor 2005). Monocytes released from the marrow into peripheral blood can mature into inflammatory (or recruited) macrophages, specialized dendritic cells, osteoclasts, and fibrocytes (Pilling et al. 2003; Gordon and Taylor 2005; Jakubzick et al. 2017). Like their neutrophil counterparts, polyP can alter monocyte migration, recruitment, and activity, as well as enhance monocyte differentiation.

Fibrocytes are a *fibro*blast-like, mono*cyte*-derived cell types that have both the properties of tissue remodeling fibroblasts and inflammatory macrophages, playing roles in innate immunity, wound healing, and fibrotic disease (Pilling et al. 2003; Reilkoff et al. 2011; Grieb and Bucala 2012). Suess et al. observed that a factor released from activated platelets enhanced fibrocyte differentiation from human PBMCs, and investigated a potential role for polyP in this differentiation inducing activity (Suess et al. 2019). Supernatant from activated platelets treated with a polyP-degrading enzyme no longer had fibrocyte-inducing activity, and PBMCs treated with synthetic platelet-sized polyP increased fibrocyte numbers (Suess et al. 2019). Platelet-derived polyP also induces fibroblast to myofibroblast differentiation, another cell type intimately involved in the wound healing process (Suess et al. 2020). These findings suggest a potential role for polyP in not only enhancing coagulation, which is critical in initiation of the wound healing process, but also by promoting the differentiation of pro-wound healing cell types.

Platelet-polyP acts as a pro-inflammatory signal on endothelial cells by increasing adhesion molecule expression (Bae et al. 2012). This activity leads to leukocyte recruitment through interactions and binding to the endothelium. Endothelial cells treated with polyP increased the expression of ICAM-1, VCAM-1, and E-selectin on the cell surface, which promotes the adhesion of monocytic THP-1 cells in vitro (Bae et al. 2012). Further, intraperitoneal injections of platelet-sized polyP in mice induced increased leukocyte binding to the vascular endothelium and subsequent migration into the peritoneal cavity (Bae et al. 2012).

Macrophages play crucial roles in host inflammatory responses, tissue repair, and homeostasis (Watanabe et al. 2019). In 2020, Ito et al. investigated the role of polyP (of approximately the size found in platelets) on the macrophage response to LPS in vitro. THP-1-derived macrophages treated with LPS increased the expression of inflammatory cytokines such as TNFα, IL-1β, and IL-6 (Ito et al. 2020). The addition of polyP significantly enhanced LPS-induced cytokine expression, while polyP had no effect in the absence of LPS (Ito et al. 2020). The results in THP-1-derived macrophages were recapitulated in human PBMCs and murine J774.1 macrophages (Ito et al. 2020). The observed effects of polyP on the macrophage response were determined to be due to a direct interaction between polyP and LPS, with polyP enhancing LPS binding to TLR4 on the surface of macrophages (Ito et al. 2020).

6.5 Host PolyP Versus Microbial PolyP

PolyP derived from platelets and mast cells appear to have largely pro-inflammatory effects on endothelial cells and leukocytes, however, less attention has been paid to the role that microbial-derived polyP plays on leukocyte function and immune responses. Work from three independent groups have reported that microbial polyP may aid in immune evasion for invading pathogens. In 2019, Terashima-Hasegawa et al. reported that polyP with an average chain length of 150 phosphate residues (larger than platelet and mast cell granular polyP) improved murine survival in an LPS-induced sepsis model by suppressing macrophage recruitment to the liver and lungs, thus preventing tissue injury (Terashima-Hasegawa et al. 2019). Macrophages treated with polyP in vitro were unresponsive to TNFα-induced chemotaxis, actin polarization, and p38 and JNK phosphorylation (Terashima-Hasegawa et al. 2019). These findings suggest that long-chain polyP can act as an immune suppressor.

In 2020, Roewe et al. investigated the role of long-chain polyP on *E. coli*-induced sepsis. Live *E. coli* were injected into mice with the presence or absence of long-chain (roughly 700 phosphate units) polyP. Long-chain polyP accelerated mortality induced by *E. coli*, while also increasing the bacterial load in the peritoneal cavity (Roewe et al. 2020). Co-injection of the *E. coli* with the polyP-degrading enzyme exopolyphosphatase (PPX) improved survival, as did injection with an *E. coli* strain deficient in the ability the make polyP (Roewe et al. 2020). The effects of long-chain polyP on leukocyte activity were also investigated, revealing that polyP suppressed monocyte maturation, suppressed monocyte and macrophage recruitment, and decreased phagocytosis of *E. coli* by monocytes, macrophages, and neutrophils (Roewe et al. 2020). To observe a more direct effect on macrophages, murine macrophages were cultured and treated with polyP and *E. coli*-derived LPS in vitro. LPS-stimulated macrophages adopt an M1 microbicidal phenotype (Roewe et al. 2020). Long-chain polyP altered the phenotype of LPS-stimulated macrophages, antagonizing M1 genes and enhancing M2 genes (Roewe et al. 2020). Taken together with the findings of Terashima-Hasegawa et al., long-chain polyP suppresses infection-induced leukocyte recruitment and anti-microbial activity. The discrepancy in the effect of polyP on sepsis survival could come from differences in model (LPS-induced vs live *E. coli*-induced) or cell types used. Roewe et al. offer another perspective, in that the size of polyP used in experimental procedures can have profound effects on the results observed, such as those seen in the survival of sepsis when $polyP_{150}$ or $polyP_{700}$ are used. Despite being identical in make-up and structure, platelet-sized or short-chain polyP (60–100 phosphates), medium-chain polyP (~150 phosphates), and microbial or long-chain polyP (300–1000 phosphates) have been identified as having different and sometimes opposing effects on the same cell type or biological activity (Smith et al. 2010; Wang et al. 2019).

Shortly after, in 2020, Rijal et al. identified further evidence that microbial polyP may support immune evasion. Extracellular polyP was found to promote the survival of *E. coli* and *M. smegmatis* phagocytosed by the eukaryotic amoeba *Dictyostelium discoideum* (Rijal et al. 2020). *Dicyostelium* feed on bacteria and utilize many of the same molecular mechanisms employed by phagocytic immune cells to track-down and engulf microbes (Dunn et al. 2017). This led the authors to speculate on the role of polyP in macrophage phagocytosis and bacterial survival. Indeed, extracellular polyP promoted the survival of *E. coli* and *M. smegmatis* engulfed by human macrophages (Rijal et al. 2020). *Mycobacterium tuberculosis* are able to survive for extended durations upon ingestion by immune cells by inhibiting phagosomal maturation and killing, leading to latent infection (Rijal et al. 2020). *M. tuberculosis* is a polyP-accumulating microbe, and *M. tuberculosis* deficient in the polyphosphate kinase PPK1 have reduced survival upon ingestion by macrophages, however the molecular mechanism by which this occurs is poorly understood (Singh et al. 2013; Tiwari et al. 2019). Here it was uncovered that the addition of recombinant PPX to co-cultures of human macrophages and *M. tuberculosis* also potentiated the killing of bacteria ingested by macrophages, leading the authors to speculate that mycobacterium may secrete polyP to improve survival upon ingestion (Rijal et al. 2020). Accordingly, *M. smegmatis* were found to accumulate extracellular polyP, and knockdown of PPK1 by CRISPRi reduced the levels of extracellular polyP accumulated and decreased viability of *M. smegmatis* upon macrophage ingestion (Rijal et al. 2020). Extracellular polyP was found to inhibit phagosome acidification and lysosomal activity of macrophages to inhibit degradation of ingested bacterium (Rijal et al. 2020). This work provides a potential link between polyP and the ability of *M. tuberculosis* to evade phagocytic degradation.

6.6 Conclusion

This review highlights that extracellular polyP modulates innate immune responses through direct and indirect effects on leukocyte activity and function. Platelet-polyP is released by activated platelets and promotes leukocyte migration, recruitment, and accumulation to the site of tissue injury or infection (Table 6.1 and Fig. 6.1). This activity occurs through multiple mechanisms, including increasing endothelial adhesion molecule expression and increasing platelet-leukocyte aggregation (Bae et al. 2012; Hou et al. 2018). In addition to directing leukocytes to areas of injury or inflammation, polyP modulates leukocyte function (Table 6.1 and Fig. 6.1). PolyP promotes NET formation and neutrophil activation, promotes monocyte maturation, and potentiates LPS-induced macrophage cytokine expression (Usui et al. 2010; Chrysanthopoulou et al. 2017; Madhi et al. 2019; Suess et al. 2019; Ito et al. 2020). The early response to tissue damage is the induction of blood coagulation to prevent

Table 6.1 Effects of polyP on leukocytes

Source	Effect	Cell type	System	References
Platelets	Promotes neutrophil-platelet-aggregate formation	Neutrophil	In vitro, in vivo	Hou et al. (2018)
Platelets	Promotes recruitment/accumulation	Neutrophil	In vivo	Hou et al. (2018)
Synthetic	Promotes recruitment/accumulation	Neutrophil	In vivo	Du et al. (2019)
Synthetic	Chemoattractant	Neutrophil	In vitro	Suess et al. (2019)
Platelets/synthetic	Promotes NET formation	Neutrophil	In vitro	Chrysanthopoulou et al. (2017)
Synthetic	Promotes autophagy	Neutrophil	In vitro	Chrysanthopoulou et al. (2017)
Platelets	Promotes NET formation	Neutrophil	In vitro	Madhi et al. (2019)
Platelets/synthetic	Promotes fibrocyte differentiation	Monocyte	In vitro	Suess et al. (2019)
Synthetic	Promotes adhesion/accumulation	Monocyte	In vitro	Bae et al. (2012)
Synthetic	Promotes recruitment/accumulation	Leukocytes	In vivo	Bae et al. (2012)
Synthetic	Potentiates LPS-induced cytokine expression	Macrophage	In vitro	Ito et al. (2020)
Synthetic	Suppresses recruitment/accumulation	Macrophage	In vitro, in vivo	Terashima-Hasegawa et al. (2019)
Microbial/synthetic	Suppresses microbicidal activity	Macrophage	In vivo	Roewe et al. (2020)
Microbial/synthetic	Suppresses maturation/differentiation	Monocyte	In vivo	Roewe et al. (2020)
Microbial/synthetic	Suppresses recruitment/accumulation	Monocyte/macrophage	In vivo	Roewe et al. (2020)
Microbial/synthetic	Inhibits phagocytosis	Monocyte/macrophage/neutrophil	In vivo	Roewe et al. (2020)
Synthetic	Antagonizes LPS response	Macrophage	In vitro	Roewe et al. (2020)
Synthetic	Suppresses microbicidal activity	Macrophage	In vitro	Rijal et al. (2020)
Synthetic	Inhibits phagosome acidification	Macrophage	In vitro	Rijal et al. (2020)
Synthetic	Inhibits lysosomal activity	Macrophage	In vitro	Rijal et al. (2020)

blood loss and recruitment of immune cells to combat potential infectious microbes. Platelets are well documented mediators of these processes, and it is now clear that platelet-polyP plays a significant role in these activities.

While platelet-polyP acts to promote host immune responses, microbial polyP can promote microbial immune evasion (Table 6.1 and Fig. 6.1). Endogenous polyP

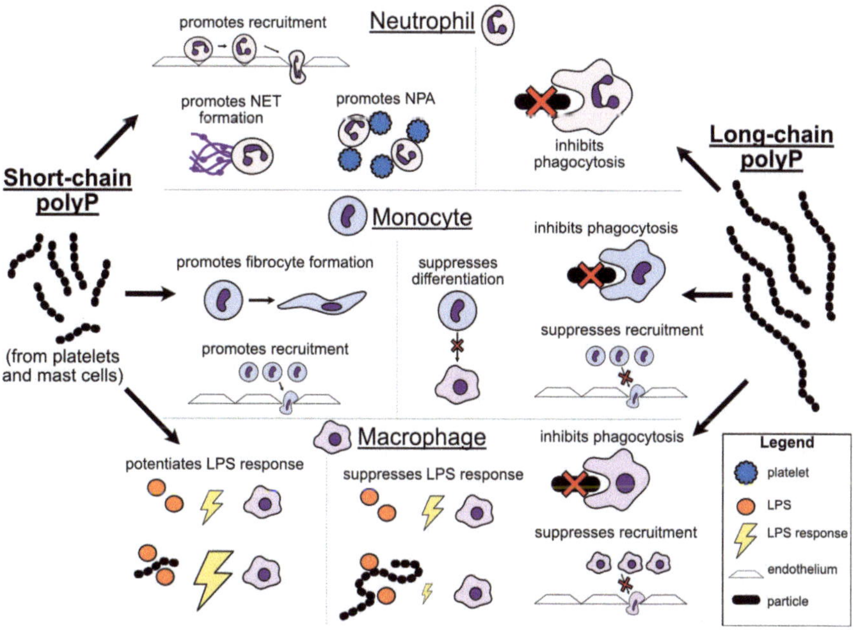

Fig. 6.1 Inflammatory and immune responses elicited by short-chain polyP and long-chain polyP

produced by infectious bacteria suppressed leukocyte recruitment to the area of infection and inhibited microbicidal activity (Roewe et al. 2020). These findings were recapitulated using synthetic long-chain polyP (Terashima-Hasegawa et al. 2019; Roewe et al. 2020). PolyP promoted the survival of bacteria upon macrophage ingestion by inhibiting phagosome acidification and lysosome activity (Rijal et al. 2020). While some microbes have been shown to secrete polyP the mechanisms by which polyP is released upon infection or ingestion is poorly understood (Suess and Gomer 2016; Neilands and Kinnby 2020; Rijal et al. 2020).

The reports detailed in this review show a clear role for extracellular polyP in mediating leukocyte function in innate immune responses (Table 6.1 and Fig. 6.1). The directly opposing roles of host-polyP and microbial polyP highlight the importance of the source and length of polyP and the effects it will induce on leukocytes. While strides have been made in understanding the role of this unique molecule in leukocyte function, many questions still remain. Due to technical limitations the importance of leukocyte intracellular polyP content, localization, and metabolism remains unknown. PolyP clearly mediates innate immune responses, however, potential roles in adaptive immunity and lymphocyte function have not been investigated. The majority of studies have focused on platelet-polyP and pathogenic microbial polyP, yet the host commensal microbiota represents a potentially large reservoir of polyP. Interactions between the mammalian immune system and commensal microbiota are known to be critical to innate and adaptive immune responses (Zheng et al. 2020). The extent of polyP accumulation of the microbiota, shifts in polyP metabolism, and potential release into the extracellular environment could all potentially influence the interplay between the microbiome and immune system.

References

Bae JS, Lee W, Rezaie AR (2012) Polyphosphate elicits pro-inflammatory responses that are counteracted by activated protein C in both cellular and animal models. J Thromb Haemost 10(6):1145–1151

Baev AY, Angelova PR, Abramov AY (2020) Inorganic polyphosphate is produced and hydrolyzed in F0F1-ATP synthase of mammalian mitochondria. Biochem J 477(8):1515–1524

Biswas I, Panicker SR, Cai X, Mehta-D'souza P, Rezaie AR (2018) Inorganic polyphosphate amplifies high mobility group box 1-mediated Von Willebrand factor release and platelet string formation on endothelial cells. Arterioscler Thromb Vasc Biol 38(8):1868–1877

Brown MR, Kornberg A (2004) Inorganic polyphosphate in the origin and survival of species. Proc Natl Acad Sci U S A 101(46):16085–16087

Cardamone C, Parente R, Feo GD, Triggiani M (2016) Mast cells as effector cells of innate immunity and regulators of adaptive immunity. Immunol Lett 178:10–14

Carney BC, Simbulan-Rosenthal CM, Gaur A, Browne BJ, Moghe M, Crooke E, Moffatt LT, Shupp JW, Rosenthal DS (2020) Inorganic polyphosphate in platelet rich plasma accelerates re-epithelialization *in vitro* and *in vivo*. Regen Ther 15:138–148

Chrysanthopoulou A, Kambas K, Stakos D, Mitroulis I, Mitsios A, Vidali V, Angelidou I, Bochenek M, Arelaki S, Arampatzioglou A, Galani IE, Skendros P, Couladouros EA, Konstantinides S, Andreakos E, Schäfer K, Ritis K (2017) Interferon lambda1/IL-29 and inorganic polyphosphate are novel regulators of neutrophil-driven thromboinflammation. J Pathol 243(1):111–122

Clark SR, Ma AC, Tavener SA, McDonald B, Goodarzi Z, Kelly MM, Patel KD, Chakrabarti S, McAvoy E, Sinclair GD, Keys EM, Allen-Vercoe E, Devinney R, Doig CJ, Green FH, Kubes P (2007) Platelet TLR4 activates neutrophil extracellular traps to ensnare bacteria in septic blood. Nat Med 13(4):463–469

Dinarvand P, Hassanian SM, Qureshi SH, Manithody C, Eissenberg JC, Yang L, Rezaie AR (2014) Polyphosphate amplifies proinflammatory responses of nuclear proteins through interaction with receptor for advanced glycation end products and P2Y1 purinergic receptor. Blood 123(6):935–945

Du F, Wang Y, Ding Z, Laschke MW, Thorlacius H (2019) Microvascular mechanisms of polyphosphate-induced neutrophil-endothelial cell interactions *in vivo*. Eur Surg Res 60(1–2): 53–62

Dunn JD, Bosmani C, Barisch C, Raykov L, Lefrançois LH, Cardenal-Muñoz E, López-Jiménez AT, Soldati T (2017) Eat prey, live: *Dictyostelium discoideum* as a model for cell-autonomous defenses. Front Immunol 8:1906

Ed Rainger G, Chimen M, Harrison MJ, Yates CM, Harrison P, Watson SP, Lordkipanidze M, Nash GB (2015) The role of platelets in the recruitment of leukocytes during vascular disease. Platelets 26(6):507–520

Finsterbusch M, Schrottmaier WC, Kral-Pointner JB, Salzmann M, Assinger A (2018) Measuring and interpreting platelet-leukocyte aggregates. Platelets 29(7):677–685

Ghosh S, Shukla D, Suman K, Lakshmi BJ, Manorama R, Kumar S, Bhandari R (2013) Inositol hexakisphosphate kinase 1 maintains hemostasis in mice by regulating platelet polyphosphate levels. Blood 122(8):1478–1486

Gordon S, Taylor PR (2005) Monocyte and macrophage heterogeneity. Nat Rev Immunol 5(12): 953–964

Grieb G, Bucala R (2012) Fibrocytes in fibrotic diseases and wound healing. Adv Wound Care (New Rochelle) 1(1):36–40
Hassanian SM, Dinarvand P, Smith SA, Rezaie AR (2015) Inorganic polyphosphate elicits pro-inflammatory responses through activation of the mammalian target of rapamycin complexes 1 and 2 in vascular endothelial cells. J Thromb Haemost 13(5):860–871
Hou Q, Liu F, Chakraborty A, Jia Y, Prasad A, Yu H, Zhao L, Ye K, Snyder SH, Xu Y, Luo HR (2018) Inhibition of IP6K1 suppresses neutrophil-mediated pulmonary damage in bacterial pneumonia. Sci Transl Med 10(435)
Iba T, Levy JH (2018) Inflammation and thrombosis: roles of neutrophils, platelets and endothelial cells and their interactions in thrombus formation during sepsis. J Thromb Haemost 16(2): 231–241
Ito T, Yamamoto S, Yamaguchi K, Sato M, Kaneko Y, Goto S, Goto Y, Narita I (2020) Inorganic polyphosphate potentiates lipopolysaccharide-induced macrophage inflammatory response. J Biol Chem 295(12):4014–4023
Jakubzick CV, Randolph GJ, Henson PM (2017) Monocyte differentiation and antigen-presenting functions. Nat Rev Immunol 17(6):349–362
Jenne CN, Kubes P (2015) Platelets in inflammation and infection. Platelets 26(4):286–292
Kolaczkowska E, Kubes P (2013) Neutrophil recruitment and function in health and inflammation. Nat Rev Immunol 13(3):159–175
Kral JB, Schrottmaier WC, Salzmann M, Assinger A (2016) Platelet interaction with innate immune cells. Transfus Med Hemother 43(2):78–88
Kulakovskaya T (2018) Inorganic polyphosphates and heavy metal resistance in microorganisms. World J Microbiol Biotechnol 34(9):139
Livermore TM, Chubb JR, Saiardi A (2016) Developmental accumulation of inorganic polyphosphate affects germination and energetic metabolism in *Dictyostelium discoideum*. Proc Natl Acad Sci U S A 113(4):996–1001
Lorenz B, Leuck J, Köhl D, Müller WEG, Schröder HC (1997) Anti-HIV-1 activity of inorganic polyphosphates. J Acquir Immune Defic Syndr Hum Retrovirol 14(2):110–118
Madhi R, Rahman M, Taha D, Linders J, Merza M, Wang Y, Mörgelin M, Thorlacius H (2019) Platelet IP6K1 regulates neutrophil extracellular trap-microparticle complex formation in acute pancreatitis. JCI Insight
Moreno-Sanchez D, Hernandez-Ruiz L, Ruiz FA, Docampo R (2012) Polyphosphate is a novel pro-inflammatory regulator of mast cells and is located in acidocalcisomes. J Biol Chem 287(34):28435–28444
Morrell CN, Aggrey AA, Chapman LM, Modjeski KL (2014) Emerging roles for platelets as immune and inflammatory cells. Blood 123(18):2759–2767
Müller WEG, Relkovic D, Ackermann M, Wang SF, Neufurth M, Paravic Radicevic A, Ushijima H, Schröder HC, Wang XH (2017) Enhancement of wound healing in normal and diabetic mice by topical application of amorphous polyphosphate. Superior effect of a host–guest composite material composed of collagen (host) and polyphosphate (guest). Polymers 9(7):300
Müller WEG, Ackermann M, Wang S, Neufurth M, Munoz-Espi R, Feng QL, Schröder HC, Wang XH (2018) Inorganic polyphosphate induces accelerated tube formation of HUVEC endothelial cells. Cell Mol Life Sci 75(1):21–32
Neilands J, Kinnby B (2020) Porphyromonas gingivalis initiates coagulation and secretes polyphosphates – a mechanism for sustaining chronic inflammation? Microb Pathog 104648
Nikel PI, Chavarria M, Martinez-Garcia E, Taylor AC, de Lorenzo V (2013) Accumulation of inorganic polyphosphate enables stress endurance and catalytic vigour in *Pseudomonas putida* KT2440. Microb Cell Factories 12:50
Pilling D, Buckley CD, Salmon M, Gomer RH (2003) Inhibition of fibrocyte differentiation by serum amyloid P. J Immunol 171(10):5537–5546
Rao NN, Kornberg A (1996) Inorganic polyphosphate supports resistance and survival of stationary-phase *Escherichia coli*. J Bacteriol 178(5):1394–1400

Rao NN, Gomez-Garcia MR, Kornberg A (2009) Inorganic polyphosphate: essential for growth and survival. Annu Rev Biochem 78:605–647

Rashid MH, Rao NN, Kornberg A (2000a) Inorganic polyphosphate is required for motility of bacterial pathogens. J Bacteriol 182(1):225–227

Rashid MH, Rumbaugh K, Passador L, Davies DG, Hamood AN, Iglewski BH, Kornberg A (2000b) Polyphosphate kinase is essential for biofilm development, quorum sensing, and virulence of *Pseudomonas aeruginosa*. Proc Natl Acad Sci U S A 97(17):9636–9641

Reilkoff RA, Bucala R, Herzog EL (2011) Fibrocytes: emerging effector cells in chronic inflammation. Nat Rev Immunol 11(6):427–435

Rijal R, Cadena LA, Smith MR, Carr JF, Gomer RH (2020) Polyphosphate is an extracellular signal that can facilitate bacterial survival in eukaryotic cells. Proc Natl Acad Sci U S A 117(50): 31923–31934

Roewe J, Stavrides G, Strueve M, Sharma A, Marini F, Mann A, Smith SA, Kaya Z, Strobl B, Mueller M, Reinhardt C, Morrissey JH, Bosmann M (2020) Bacterial polyphosphates interfere with the innate host defense to infection. Nat Commun 11(1):4035

Rossaint J, Margraf A, Zarbock A (2018) Role of platelets in leukocyte recruitment and resolution of inflammation. Front Immunol 9:2712

Ruiz FA, Lea CR, Oldfield E, Docampo R (2004) Human platelet dense granules contain polyphosphate and are similar to acidocalcisomes of bacteria and unicellular eukaryotes. J Biol Chem 279(43):44250–44257

Singh R, Singh M, Arora G, Kumar S, Tiwari P, Kidwai S (2013) Polyphosphate deficiency in *Mycobacterium tuberculosis* is associated with enhanced drug susceptibility and impaired growth in Guinea pigs. J Bacteriol 195(12):2839–2851

Singh M, Tiwari P, Arora G, Agarwal S, Kidwai S, Singh R (2016) Establishing virulence associated polyphosphate kinase 2 as a drug target for *mycobacterium tuberculosis*. Sci Rep 6:26900

Smith SA, Mutch NJ, Baskar D, Rohloff P, Docampo R, Morrissey JH (2006) Polyphosphate modulates blood coagulation and fibrinolysis. Proc Natl Acad Sci U S A 103(4):903–908

Smith SA, Choi SH, Davis-Harrison R, Huyck J, Boettcher J, Rienstra CM, Morrissey JH (2010) Polyphosphate exerts differential effects on blood clotting, depending on polymer size. Blood 116(20):4353–4359

Suess PM, Gomer RH (2016) Extracellular polyphosphate inhibits proliferation in an autocrine negative feedback loop in *Dictyostelium discoideum*. J Biol Chem 291(38):20260–20269

Suess PM, Watson J, Chen W, Gomer RH (2017) Extracellular polyphosphate signals through Ras and Akt to prime *Dictyostelium discoideum* cells for development. J Cell Sci 130(14): 2394–2404

Suess PM, Chinea LE, Pilling D, Gomer RH (2019) Extracellular polyphosphate promotes macrophage and fibrocyte differentiation, inhibits leukocyte proliferation, and acts as a chemotactic agent for neutrophils. J Immunol 203(2):493–499

Suess PM, Smith SA, Morrissey JH (2020) Platelet polyphosphate induces fibroblast chemotaxis and myofibroblast differentiation. J Thromb Haemost 18(11):3043–3052

Terashima-Hasegawa M, Ashino T, Kawazoe Y, Shiba T, Manabe A, Numazawa S (2019) Inorganic polyphosphate protects against lipopolysaccharide-induced lethality and tissue injury through regulation of macrophage recruitment. Biochem Pharmacol 159:96–105

Tinsley CR, Manjula BN, Gotschlich EC (1993) Purification and characterization of polyphosphate kinase from *Neisseria meningitidis*. Infect Immun 61(9):3703–3710

Tiwari P, Gosain TP, Singh M, Sankhe GD, Arora G, Kidwai S, Agarwal S, Chugh S, Saini DK, Singh R (2019) Inorganic polyphosphate accumulation suppresses the dormancy response and virulence in *Mycobacterium tuberculosis*. J Biol Chem 294(28):10819–10832

Usui Y, Uematsu T, Uchihashi T, Takahashi M, Takahashi M, Ishizuka M, Doto R, Tanaka H, Komazaki Y, Osawa M, Yamada K, Yamaoka M, Furusawa K (2010) Inorganic polyphosphate induces osteoblastic differentiation. J Dent Res 89(5):504–509

Wang Y, Ivanov I, Smith SA, Gailani D, Morrissey JH (2019) Polyphosphate, Zn^{2+} and high molecular weight kininogen modulate individual reactions of the contact pathway of blood clotting. J Thromb Haemost 17(12):2131–2140

Watanabe S, Alexander M, Misharin AV, Budinger GRS (2019) The role of macrophages in the resolution of inflammation. J Clin Invest 129(7):2619–2628

Zheng D, Liwinski T, Elinav E (2020) Interaction between microbiota and immunity in health and disease. Cell Res 30(6):492–506

Chapter 7
Polyphosphate in Antiviral Protection: A Polyanionic Inorganic Polymer in the Fight Against Coronavirus SARS-CoV-2 Infection

Werner E. G. Müller, Xiaohong Wang, Meik Neufurth, and Heinz C. Schröder

Abstract Polyanions as polymers carrying multiple negative charges have been extensively studied with regard to their potential antiviral activity. Most studies to date focused on organic polyanionic polymers, both natural and synthetic. The inorganic polymer, polyphosphate (polyP), despite the ubiquitous presence of this molecule from bacteria to man, has attracted much less attention. More recently, and accelerated by the search for potential antiviral agents in the fight against the pandemic caused by the coronavirus SARS-CoV-2, it turned out that polyP disrupts the first step of the viral replication cycle, the interaction of the proteins in the virus envelope and in the cell membrane that are involved in the docking process of the virus with the target host cell. Experiments on a molecular level using the receptor-binding domain (RBD) of the SARS-CoV-2 spike protein and the cellular angiotensin converting enzyme 2 (ACE2) receptor revealed that polyP strongly inhibits the binding reaction through an electrostatic interaction between the negatively charged centers of the polyP molecule and a cationic groove, which is formed by positively charged amino acids on the RBD surface. In addition, it was found that polyP, due to its morphogenetic and energy delivering activities, enhances the antiviral host innate immunity defense of the respiratory epithelium. The underlying mechanisms and envisaged application of polyP in the therapy and prevention of COVID-19 are discussed.

Keywords SARS-CoV-2 · COVID-19 · Polyphosphate · Antiviral · Polyanion · Spike protein · Receptor-binding domain · ACE2 · Cytokines · Mucus

W. E. G. Müller (✉) · X. Wang · M. Neufurth · H. C. Schröder
ERC Advanced Investigator Group, Institute for Physiological Chemistry, University Medical Center of the Johannes Gutenberg University, Mainz, Germany
e-mail: wmueller@uni-mainz.de; wang013@uni-mainz.de; hschroed@uni-mainz.de

W. E. G. Müller et al. (eds.), *Inorganic Polyphosphates*, Progress in Molecular and Subcellular Biology 61, https://doi.org/10.1007/978-3-031-01237-2_7

7.1 Introduction

The pandemic caused by the new coronavirus, the severe acute respiratory syndrome coronavirus 2 (SARS-CoV-2), has developed into a global problem not only for health, but also for the economy and society with as yet unforeseeable consequences. This virus, which is transmitted by small droplets when coughing or sneezing, as well as by aerosols that are produced during normal breathing or speaking, can cause serious illness with pneumonia and acute respiratory distress syndrome and ultimately affect many organs. The mortality of the disease (Coronavirus Disease-19, COVID-19) strongly depends on the age of the patient and is higher in men than in women. There is also an increased risk for patients with preexisting diseases, such as diabetes, hypertension, or cardiovascular diseases.

The causative agent of COVID-19, SARS-CoV-2, is an RNA virus with a size of 80–120 nm, which is surrounded by a lipid envelope in which the viral spike (S) proteins are anchored, via which the virus docks to the ACE2 (angiotensin converting enzyme 2) receptor on the host cell (Hoffmann et al. 2020a).

In 2020/2021, several vaccines have been developed and approved, but there is still a lack of them, especially in Third World countries and the number of anti-vaccinationist in countries without mandatory vaccination is relatively high. In particular, due to the emergence of new mutants/variants of the virus, there is a well-founded fear that existing vaccines can become ineffective, leading to the occurrence of new epi- or pandemics.

Efficient drug therapy for COVID-19 patients is not yet available (Gil et al. 2020; Hussman 2020). Drug repurposing strategies did not lead to convincing results (Kifle et al. 2021). Thus, therapy for COVID-19 remains a great challenge. In addition, there is increasing recognition that it is important also to develop effective drugs that can be used to treat or prevent infection by future, as yet unknown zoonotic viruses.

One polyanionic polymer has recently attracted attention as a potential medication for therapy of COVID-19 patients, but also for the prevention of transmission of SARS-CoV-2. This polymer, inorganic polyphosphate (polyP) has the advantage that it is physiological, obviously present in all body cells as well as extracellular fluids, and is without any toxicity (Müller et al. 2019). In humans and other vertebrates, polyP is synthesized in the mitochondria and probably the acidocalcisomes (Docampo et al. 2005; Müller et al. 2019). ATP, which is produced by F_1F_O ATP synthase and released from the mitochondria *via* the ADP/ATP translocase and the voltage-dependent anion channels (VDAC) in the inner and outer mitochondrial membranes, has been proposed to be converted to polyP by the polyP polymerase of the acidocalcisomal vacuolar transporter chaperone (Vtc) complex (Müller et al. 2019), which is active at least in yeast (Gerasimaitė et al. 2014). This complex also channels the polyP into the acidocalcisomes, which serve as storage organelles of the polymer. Large amounts of acidocalcisomes (dense granules) are present in the platelets, which are split off from their precursors, the megakaryocytes (Ambrosio and Di Pietro 2017) and, after activation, release the polyP (Morrissey et al. 2012).

Recent results revealed that polyP can interact with various (extracellular) pathogenic proteins, e.g., Alzheimer's amyloid-β (Aβ) peptide or tau protein, as well as α-synuclein involved in Parkinson's disease (Lempart et al. 2019; Wickramasinghe et al. 2019; Xie and Jakob 2019), but also infectious virus. In previous studies, we discovered that polyP has a strong antiviral effect on HIV-1 (Lorenz et al. 1997). Based on this study and results showing that polyelectrolytes, in general, are able to efficiently block binding of viruses to the target cells, a study on the interaction of polyP with the SARS-CoV-2 spike protein through which the virus docks to its target host cells and the possible beneficial effects of the polymer with regard to therapy or prevention COVID-19 has been started. It turned out that polyP not only efficiently blocks the virus–receptor interaction, but is also capable of significantly enhancing the innate immune defense of the host against the virus.

7.2 Polyanions as Antiviral Agents

Many polyanions, being polymeric compounds containing multiple negative charges, have been shown to be potent antiviral agents in vitro (Bianculli et al. 2020). This class of antiviral molecules comprises both biopolymers and synthetic polymers as well as derivatives of the natural molecules. A common advantageous property of these polyanions, which distinguishes them at least from most other antiviral agents, is that they are not only able to inhibit the attachment of the virus to the target host cell membrane, but also have a comparatively broad-spectrum antiviral activity and the ability to suppress syncytium formation (Lüscher-Mattli 2000). It has also been reported that they can interfere with the fusion process of viruses with the host cell membranes (Lüscher-Mattli 2000).

7.2.1 Bioorganic Polyanions

Most of the polyanions identified to date to exhibit antiviral activity are sulfated polysaccharides or synthetic sulfated polymers (Bianculli et al. 2020). Among the sulfated polysaccharides, the anti-HIV activity of dextran sulfate is well investigated (Ueno and Kuno 1987; Nakashima et al. 1987; Mitsuya et al. 1988). Dextran sulfate binds to the HIV gp41 (Gordon et al. 1995). This polymer, as well as pentosan polysulfate, has been shown to suppress the fusion of influenza virus, vesicular stomatitis virus, and rabies virus in vitro (Lüscher-Mattli and Glück 1990; Lüscher-Mattli et al. 1993), most likely by binding to the viral fusion peptide, resulting in inactivating of the protein. It has been proposed that this mechanism is due to electrostatic interactions between the polyanionic polymer and positively charged amino acids of the fusion peptide (Hiti et al. 1981). Further examples of polyanionic biomolecules that are antivirally active are polysulfated polyxylan, which inhibits HIV-1 and -2 replication in macrophages (von Briesen et al. 1990), as well as the polysaccharide sulphoevernan, which has been found to potently inhibit HIV-1 infection by binding to viral envelope gp120 protein (Weiler et al. 1990).

Besides these bioorganic polyanions, there are several synthetic organic polyanions that exhibit antiviral activity. Examples are sulfated polyvinylalcohol (PVAS) and the copolymer of acrylic acid with vinyl alcohol sulfate (PAVAS), which are antivirally active against a number of viruses including herpes simplex viruses, vesicular stomatitis virus, human cytomegalovirus, and HIV-1/-2 (Schols et al. 1990; Hosoya et al. 1991), as well as sulfonic acid polymers such as poly (vinylsulfonic acid) (PVS), poly(anetholesulfonic acid) (PAS), poly(4-styrene sulfonic acid) (PSS), and poly(2-acrylamido-2-methyl-1-propanesulfonic acid) (PAMPS); the latter polymer, in in vitro experiments, has been shown to act inhibitory on syncytia formation and replication of HIV-1 and -2 (Mohan et al. 1992; Tan et al. 1993).

An extensively studied organic polyanion is heparin, a glycosaminoglycan composed of variably sulfated disaccharide units (Mulloy et al. 2016). This polymer is widely used because of its effect on blood coagulation. More recently, this polymer also attracted interest with regard to its possible application for treatment of COVID-19 patients showing hypercoagulability (Gozzo et al. 2020). Further studies showed that heparin interacts with the receptor binding domain (RBD) of the SARS-CoV-2 spike protein (Mycroft-West et al. 2020a). Exposure of the viral RBD to heparin was found to lead to a conformational change of the RBD (Mycroft-West et al. 2020a, 2020c). However, in later studies it has been demonstrated that heparin does not inhibit but, at higher concentrations, rather increases the binding propensity of the RBD to the ACE2 receptor (Müller et al. 2020b).

7.2.2 *Bioinorganic Polyanion: Polyphosphate (PolyP)*

Inorganic polyphosphates (polyP) are the major group of physiological inorganic polymers that have attracted increasing medical interest in recent years. These "biomedical inorganic polymers" (Müller et al. 2013) can be present in different size classes that differ in their physicochemical and biological properties. The chain lengths of polyP released from human platelets into the circulating blood are in the range from 50 to 100 P_i units (Morrissey et al. 2012). At these chain lengths, less than 100 P_i residues, the sodium salt of polyP (Na-polyP) is readily soluble in water (Katchman and Smith 1958; Van Wazer 1958). Long-chain polyP molecules (~1000 P_i units) are mainly found in bacteria (Kulaev et al. 2004), with some exceptions; e.g., as long-chain polyP (~200 – 1000 P_i units) of nanoparticles bound to the platelet surface (Verhoef et al. 2017). The solubility of polyP decreases with the chain length of the polymer, but also short-chain polyP can be present in an insoluble form, if prepared as amorphous polyP nano/microparticles of the salts of the polymer with divalent cations (Ca^{2+}, Mg^{2+}, Sr^{2+}, Zn^{2+}, and others; Müller et al. 2018a; Schröder et al. 2020). The physiological effects of short-chain polyP found in humans and bacterial long-chain polyP are quite different, as will be described in this chapter.

The biological properties of long-chain polyP from bacteria and yeast and of human short-chain polyP have been summarized in several reviews and monographs (e.g., Kornberg et al. 1999; Kulaev et al. 2004; Baker et al. 2018; Xie and Jakob 2019; Müller et al. 2019). In this chapter, we will mainly focus on the antiviral properties of short-chain polyP.

The first study showing that the inorganic polymer, polyP, exhibits antiviral/anti-HIV activity in vitro, was published in 1997 (Lorenz et al. 1997). It was found that polyP of different chain lengths (15, 34, and 91 P_i residues) inhibits the replication of human immunodeficiency virus type 1 (HIV-1) as well as HIV-1-induced syncytium formation; sodium tripolyphosphate was ineffective. The results of competition experiments indicated that polyP interferes with the binding of the virus to the cellular CD4 receptor (Lorenz et al. 1997).

7.2.3 Beneficial Properties of PolyP

PolyP is provided with a unique set of advantageous properties that make this molecule highly suitable for potential application as a drug in antiviral therapy.

First, it is a strong polyanion. At physiological pH, each internal P_i unit of polyP has a negative charge; the terminal P_i units can even carry two negative charges. The corresponding dissociation constants are $pK_1 = 2.2$ for the internal and the first terminal OH group and $pK_2 = 7.2$ for the second terminal OH (Lippmann 1951; Kirk-Othmer Encyclopedia of Chemical Technology 1991). In alkaline solutions, the OH groups are nearly completely dissociated. The charge density of the polyP chain thereby depends on the intramolecular electrostatic repulsions within the polymer chain. From the length of the P–O bond (1.65 Å) a distance between the negative charges of two adjacent PO_4 tetrahedra of 2.61 Å has been calculated (Cini and Ball 2014). Polyanionic polymers, as described above, are efficient inhibitors of the attachment of viruses to their cellular receptor molecules.

Second, polyP is a flexible molecule, which is able to enwrap and thereby mask proteins. The bond angles of the P–O–P linkages (130°) and of the O–P–O bonds (102°) in the polyP molecule can vary in dependence on the counterions; e.g., the P–O–P bond angle can range between ~120° and ~180° (Majling and Hanic 1980). In addition, the energy barrier for the rotation around the P-O-P bond only amounts to 1.0 kcal mol^{-1} (Semlyen and Flory 1966). This low torsional barrier surely contributes to the high flexibility of the linear polyP molecules (Yoshida 1955), allowing the polymer to adapt to and, e.g., to affect the assembly of pathogenic proteins, such as Alzheimer's Aβ peptide.

Third, polyP is a stable in aqueous solution at room temperature and neutral, physiological pH, although the standard Gibbs free energy (ΔG^0) of hydrolysis of the P–O–P bond is high (like in ATP with −30.5 kJ mol^{-1}) and the chemical equilibrium is almost totally on the side of the products of the hydrolytic reaction, the monomeric orthophosphates. The stability of polyP even increases under alkaline conditions. Only in acidic solution, hydrolysis readily proceeds (Van Wazer 1958; Katchman and Smith 1958; Kulaev et al. 2004; Rao et al. 2009). The kinetic stability of polyP at neutral or alkaline pH, despite its thermodynamic instability, is due to the high activation energy (E_a) for the hydrolytic cleavage of the P-O-P bond, like in ATP, which is most likely caused by the negative charges of the P_i units of this strong polyanion, which impede the nucleophilic attack of water molecules (Lippmann 1951; Thilo and Wicker 1957; Van Wazer 1958; Westheimer 1987). Therefore, the hydrolytic degradation of polyP requires the presence of an enzyme to reduce the activation energy, or an acidic pH (Kulaev et al. 2004; Rao et al. 2009). The acid-catalyzed hydrolysis starts from the ends of the polymer and involves the formation of a pentavalent intermediate (De Jager and Heyns 1998a, b). The protonation of the double-bonded oxygen of the terminal P_i unit results in a decrease of E_a (from >104 kJ mol^{-1} to 57 kJ mol^{-1}; De Jager and Heyns 1998a), which facilitates the formation of the transition state by the attacking water molecules. The enzymatic cleavage of the P–O–P bond, on the other hand, has been proposed to involve the formation of a metaphosphate intermediate (Müller et al. 2019). The responsible enzyme, the alkaline phosphatase (ALP), can transfer this phosphate species to a phosphate acceptor molecule such as AMP, enabling the formation of ADP, and finally ATP by the combined action of ALP and adenylate kinase (ADK; see below).

Fourth, polyP shows morphogenetic activity. PolyP is able to induce the expression of certain genes, including genes that are involved in crucial steps in cell differentiation (for reviews, see Wang et al. 2014; Müller et al. 2018a; Schröder et al. 2019) as well as genes that are involved in the response of cells to stresses and pathogenic agents or repair/regeneration processes.

Fifth, polyP is able to form a coacervate that can encase viruses or other pathogenic agents. PolyP, in the physiologically relevant chain length range, can be prepared either in the form of amorphous nano/microparticles or as a coacervate. The nano/microparticles are fabricated from the soluble polyP (the sodium salt, Na-polyP; usually polyP with a chain length of 40 P_i units is used) in the presence of an excess amount of divalent cations, e.g., Ca^{2+}, Mg^{2+}, and Zn^{2+} ions, under alkaline conditions (pH 10). The resulting particles, e.g., Ca-polyP nano/microparticles, have a negative zeta (ζ) potential due to a charge separation in aqueous solution that prevents the aggregation of the particles and contributes to their high stability. The polyP coacervate is obtained from soluble polyP and divalent cations at neutral pH, in particular, in the presence of proteins/peptides. This coacervate is the biologically/morphogenetically active form of the polymer that enables cells incorporated in this phase to proliferate and differentiate (Müller et al. 2018b), but can also be used to scavenge virus particles or other pathogens that can be embedded in the coacervate as described below. In therapeutic applications, the stable nano/microparticles can serve as a storage form of the polymer that, after contact with protein-containing body fluids, is activated by transformation in the coacervate phase (Müller et al. 2018b).

Sixth, polyP can provide metabolic energy, which is needed to strengthen the resistance of cells and tissues against damage by bacteria, viruses, and other pathogens. The ALP, which catalyzes the exergonic hydrolytic cleavage of the phosphoanhydride bonds in polyP (like in ATP, $\Delta G^0 \approx -30.5\ \text{kJ mol}^{-1}$ per phosphoanhydride bond), can also act as a phosphotransferase by transferring the metaphosphate intermediate formed during the reaction to AMP under formation of ADP (for a review, see Müller et al. 2019). The ALP thereby acts in cooperation with a second enzyme, bound to the outer cell membrane, the ADK, which generates ATP from ADP by interconversion of two molecules of ADP to ATP and AMP (Müller et al. 2017c). Thereby, the chemical energy stored in polyP is converted into metabolically useful energy.

A disadvantage of polyanions, in general, could be a low absorption if given orally as found in clinical trials (Abrams et al. 1989; Lorentsen et al. 1989). However, the nasal route might be an alternative, as used with polyP (Schepler et al. 2021).

7.3 The Virus: SARS-CoV-2

The coronavirus SARS-CoV-2 is a single-stranded positive-sense RNA betacoronavirus (Lu et al. 2020). The prominent protein at the outer surface of the lipid envelope that surrounds the virus particle is the spike (S) glycoprotein *via* which the virion binds to its host cell receptor, the angiotensin converting enzyme 2 (ACE2) (Hoffmann et al. 2020a). In addition to the spike S-protein and three further structural proteins, envelope (E), membrane (M), and nucleocapsid (N) proteins, the virus genome encodes a series of nonstructural proteins, such as RNA-dependent RNA polymerase (RdRp), helicase (Hel), exonuclease (ExoN), papain-like protease (PLpro), and 3C-like protease (3CLpro), which are involved in the viral replication cycle (Chan et al. 2020). Many of these proteins are potential targets for antiviral compounds (Huang et al. 2020; Tang et al. 2020; Zhang et al. 2020).

7.4 The Viral Life Cycle: Attack on the Target Cell

The replication cycle of most viruses, such as that of SARS-CoV-2, can be roughly divided into the phases adsorption, fusion, uncoating, replication, transcription of viral mRNA and translation of viral proteins, assembly, maturation, and release. After recognition and binding of the virus to the receptor on the target cell membrane, the entry of the virus into the host cell can occur either via endocytosis or membrane fusion. The expression of the viral genome often leads to the synthesis of viral polyprotein that must be proteolytically cleaved into the mature viral proteins.

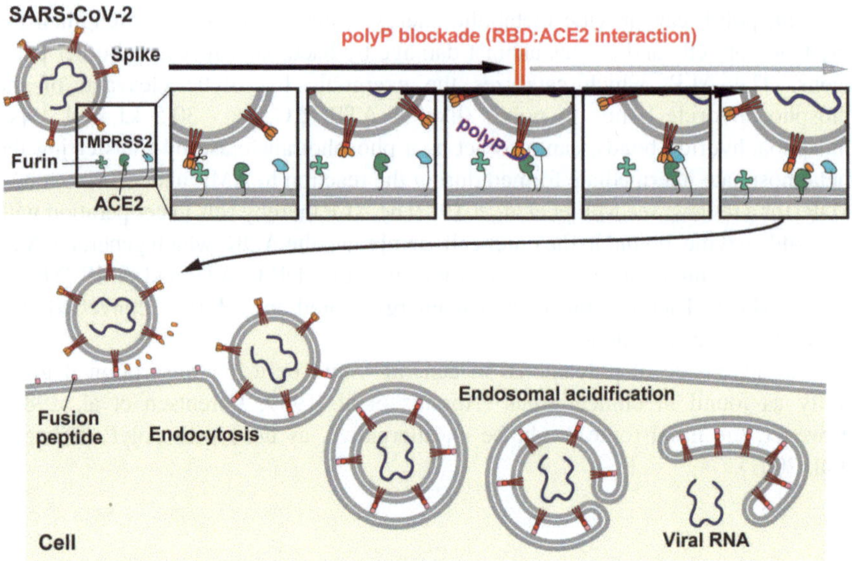

Fig. 7.1 First steps of the replication cycle of SARS-CoV-2. Top: Attachment of the virus particles to the host cell receptor ACE2. The homotrimeric spike S-proteins that protrude from the viral envelope bind to the cellular receptors via their receptor binding domains (RBD) at the tips of their S1-subunits. The receptor binding requires a prior change of the conformation of the RBD from a closed down state to an open up state. The RBD:ACE2 interaction is accompanied by a proteolytic cleavage of the spike protein at the S1/S2 boundary of the S1 and S2 subunits by the cellular proteases TMPRSS2 and furin, or by the lysosomal protease cathepsin L after endocytosis. Thereby, the S1-subunit is liberated or remains attached to the receptor. Bottom: After cleavage, the fusion peptide of the remaining S2-subunit, which is still anchored in the virus membrane, becomes inserted into the host cell membrane. This process enables the fusion of the virus and host cell membranes, allowing the release of the virus RNA into the host cell

Here we are only looking at the first steps in the SARS-CoV-2 replication cycle, which begins with the binding of the spike S-protein to the cellular ACE2 receptor, followed by the entry of the virus into the host cell either through endocytosis or membrane fusion (Fig. 7.1).

7.4.1 The Spike Protein: The Prominent Protein at the Virus Surface

The spike (S) glycoprotein, which is exposed on the virus surface, is the main protein involved in the first steps of viral replication cycle. This trimeric protein consists of a short intracellular *C*-terminal segment, a transmembrane domain, and a large extracellular *N*-terminal segment (Walls et al. 2020). Each monomer of this homotrimer comprises two subunits, the S1 subunit, which is responsible for binding to the host cell ACE-2 receptor (Lan et al. 2020), and the S2 subunit, which mediates fusion of the viral and cellular membranes (Ou et al. 2020). The S1 subunit carries the RBD that binds to the extracellular peptidase domain of ACE2 (Lu et al. 2020; Dong et al. 2020; Yan et al. 2020). The S2 subunit has two hydrophobic repeat regions, HR1 and HR2, and the fusion peptide, which are involved in the fusion process (Shang et al. 2020; Walls et al. 2020; Xia et al. 2020). During the RBD:ACE2 interaction, a cleavage site of S2 is exposed and the S1-S2 heterodimers are cleaved by host cell proteases into their two functional subunits, S1 and S2 (Ou et al. 2020).

Two cleavage sites are present in the S-protein: the S1/S2 site at the boundary of the S1 and S2 subunits, and the S2′ site, which is located close to the fusion peptide (Hulswit et al. 2016; Hoffmann et al. 2018; Millet and Whittaker 2018). The main cleavage site is the S1/S2 site, which comprises a cluster of basic amino acids with a total of 3 Arg residues (Walls et al. 2020; Wrapp et al. 2020). A series of host cell proteases are involved in the proteolytic cleavage of S-protein at the S1/S2 site, among them the serine proteases TMPRSS2, furin, and cathepsin L (Coutard et al. 2020; Hoffmann et al. 2020b,c). TMPRSS2 is located at the plasma membrane of the host cell, while cathepsin L is a lysosomal enzyme, which is optimally active under acidic pH conditions and most likely involved in the cleavage reaction in the endosomes after endocytic uptake of the virus and fusion with lysosomes (Liu et al. 2020). TMPRSS2 needs a neutral milieu to be catalytically active (Meyer et al. 2013). Only after proteolytic activation, fusion of the viral and cellular membranes and release of virus RNA into the host cell can occur (Hoffmann et al. 2020b).

7.4.2 The Receptor Binding Domain: The Spear Head of Spike Protein

The RBD is located at the tip of the SARS-CoV-2 S-protein (Tai et al. 2020; Ou et al. 2020); Fig. 7.2a. The protein consists of 218 amino acids (molecular mass, 24.5 kDa) and has a positive charge due to the higher number of basic amino acids compared to acidic amino acids. In total, the RBD contains 11 arginine (Arg), 10 lysine (Lys), and 1 histidine (His) residues, while the sum of the acidic amino acids, aspartate (Asp) and glutamate (Glu) amounts to 15 (Neufurth et al. 2020; Müller et al. 2020a); the calculated isoelectric point (pI) is 8.9 (Tai et al. 2020). In contrast, the host cell receptor of SARS-CoV-2, the ACE2, is a negatively charged protein (Zhang et al. 2020).

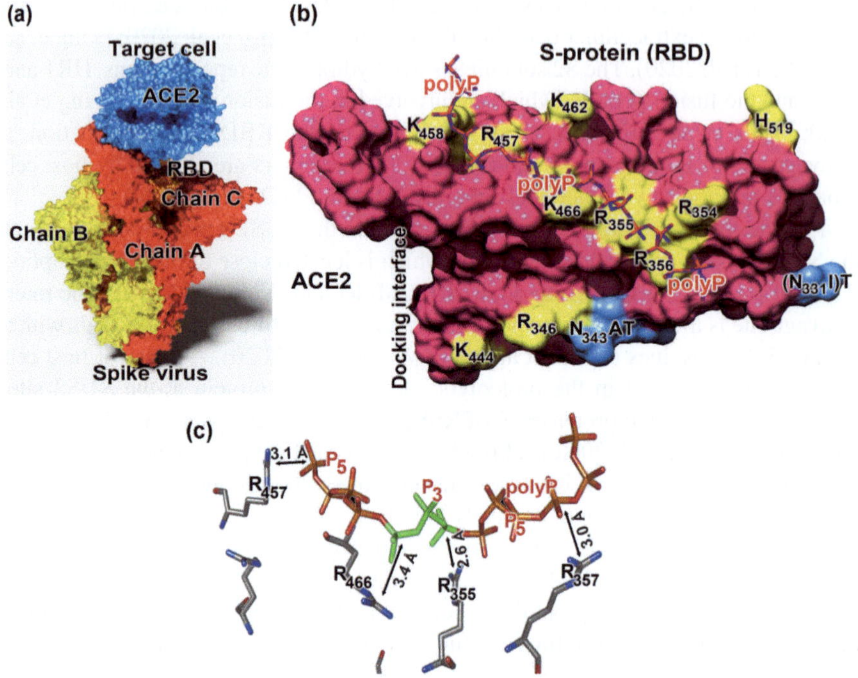

Fig. 7.2 (**a**) Scheme of the trimeric SARS CoV-2 spike protein with the chains A, B, and C and the position of the RBD, bound to ACE2 (only one RBD is shown). Adapted with permission from Schepler et al. (2021). Copyright 2021, Ivyspring International Publisher, Sydney, Australia. (**b**) Model of the RBD with the attached polyP. The basic amino acids are labeled in yellow and the nonpolar and acidic amino acids are given in pink. The amino acids with the putative *N*-glycosylation sites are shown in blue. The positions of the amino acids that are involved in polyP binding are indicated. The side of the docking interface to ACE2 is on the left. Adapted with permission from Neufurth et al. (2020). Copyright 2020, Elsevier Inc. (**c**) Distances between the arginine guanidinium groups of the RBD cationic groove and the phosphate groups of the polyP (in Ångström, Å) bound to the RBD. Adapted with permission from Müller et al. (2020a). Copyright 2020, Royal Society of Chemistry

7.4.3 The Host Cell Receptor: Angiotensin Converting Enzyme 2 (ACE2)

The cellular ACE2 protein, which is used as receptor by SARS-CoV-2, is a dimeric zinc-containing metalloprotein. This protein is mainly present in the respiratory tract. It is expressed in alveolar epithelial type II cells; it can also be found in the vascular endothelium of lung, heart, intestine, and kidney (Lu et al. 2020; Zhang et al. 2020). ACE2 has enzymatic activity. It acts, like its homolog ACE, as a carboxypeptidase. In contrast to ACE, which converts angiotensin I (10 amino acids) to angiotensin II (8 amino acids), ACE2 cleaves angiotensin I under the formation of angiotensin 1–9

(9 amino acids); in addition, it cleaves angiotensin II to angiotensin 1–7 (7 amino acids) (Donoghue et al. 2000); see Sect. 7.5.

7.4.4 The Spike Protein:ACE2 Interaction

The spike protein of SARS-CoV-2 binds to the host cell ACE2 receptor through the RBD domain of its trimeric S1 subunit (Wrapp et al. 2020). This binding is strong; the dissociation constant amounts to $K_D = 14.7$ nM (Wrapp et al. 2020). The RBD: ACE2 interaction is associated with a conformational change of the S1 subunit. The RBD exists in two conformations: a closed down state, which is unable to efficiently bind to the receptor and an open up conformation, which is accessible to the receptor (Walls et al. 2020; Wrapp et al. 2020). Only in the open-up conformation, the docking reaction occurs.

The RBD:ACE2 interaction and internalization of the virus have been proposed to be facilitated by heparan sulfate, which is present on the host cell surface (Clausen et al. 2020; Kim et al. 2020). This glycosaminoglycan molecule has been proposed to interact with the S1/S2 proteolytic cleavage site of the spike protein, which is characterized by a basic amino acid stretch, supporting the virus entry into the host cell (Kim et al. 2020).

7.4.5 The Receptor Binding Domain and Its Cationic Groove: Interaction with PolyP

A unique feature of the RBD is the presence of a particular arrangement of the basic amino acids Arg and Lys that form a cationic groove on the surface of the protein (Lan et al. 2020; Ou et al. 2020); Fig. 7.2b. This specific amino acid cluster on the RBD surface comprises six of the nine basic amino acids (five Arg and four Lys residues), which are exposed on the RBD surface. At physiological pH, the side chains of the clustered Arg (pI, 10.8) and Lys (pI, 9.5) residues that form this pattern are positively charged. The spaces between these amino acids reach from 3.8 Å to 5 Å. This specific arrangement of basic amino acids allows the RBD to interact with polyanionic polymers, such as the inorganic polyP (Neufurth et al. 2020). A similar interaction has been proposed for the organic polyanionic molecule, heparin (Mycroft-West et al. 2020b).

Modelling studies revealed that the polymeric polyP with its negatively charged phosphate residues perfectly fits to the spatial arrangement of the basic amino acids that form the cationic groove at the RBD surface (Neufurth et al. 2020; Müller et al. 2020a), suggesting this polymer exhibits a strong propensity to associate with the viral RBD surface. The spacing between the phosphate units of the polyP is ~3 Å, and the distances between the guanidinium groups of the Arg residues and the phosphate groups range between 2.6 and 3.4 Å (Fig. 7.2c).

7.4.6 The Mechanism of Interaction

The binding of polyP to the RBD is most likely due to electrostatic interactions (Vicenzi et al. 2004) with the Arg and Lys residues on the RBD surface. The terminal guanidinium group of the Arg side chain, consisting of three amino groups bound to a central carbon, assumes a planar, symmetric structure with a resonance-stabilized positive charge (Fig. 7.3a). The electrostatic forces can be increased by adjacent acidic amino acids (Suckau et al. 1992). Indeed, the cationic groove of the RBD is traversed by a pair of two acidic amino acids, Asp467 and Glu465, that are located at the two cationic Arg residues Arg457 and Arg466 (Müller et al. 2020a). This specific amino acid arrangement forms an intramolecular proton transfer system that facilitates the formation of covalent linkages of the Arg guanidinium group with certain additives (Qian et al. 2004; Suckau et al. 1992; Mitchell et al. 1992).

Calculations using the algorithm of Kyte and Doolittle (1982) revealed a hydropathy index of −4.5 for this amino acid stretch formed by the clustered amino acids Arg (index −4.5) and Lys (index −3.9) (Müller et al. 2020a). These results were strengthened by calculations of the electrostatic potential by applying the Poisson-Boltzmann equation (Xie et al. 2020). The positively charged Arg-/Lys-rich stretch and negatively charged Asp/Glu crossing area match the prediction of the modelling experiments (Müller et al. 2020a).

The RBD-bound polyP most likely assumes a staggered conformation (Fig. 7.2c). This conformation is also shown by the monovalent Na^+ salt of polyP while the divalent Ca^{2+} salt of polyP prefers an eclipsed conformation (McCarthy et al. 2002).

These results showing interaction of SARS-CoV-2 with polyP have recently been confirmed also with long-chain polyP ($polyP_{120}$; Ferrucci et al. 2021). In addition, this group demonstrated that polyP is able to interact with the viral RNA-dependent RNA polymerase (RdRp), resulting in a disturbance of viral subgenomic RNA synthesis.

It should be mentioned that in addition to SARS-CoV-2, Arg-rich sequences that can penetrate cell membranes are also found in other viral peptides/proteins. An example is the HIV-1 Tat protein that contains a basic stretch formed by six Arg, two Lys, and one polar/non-charged amino acids (glutamine, Gln) (Gasparini et al. 2015); Fig. 7.3b. After crossing the cell membrane, Tat is able to bind to the TAR sequence of the HIV-1 mRNA, which forms a double-stranded stem-loop structure (Müller et al. 1990).

It is assumed that the guanidinium groups present in cell-penetrating Arg-rich peptides such as HIV-1 Tat protein interact with the membrane lipid phosphate groups, resulting in a reorganization of the lipids (Pantos et al. 2008). The membrane passage of these peptides has been proposed to involve the generation of negative Gaussian curvature at the membrane (Schmidt et al. 2010; Mishra et al. 2011) or the

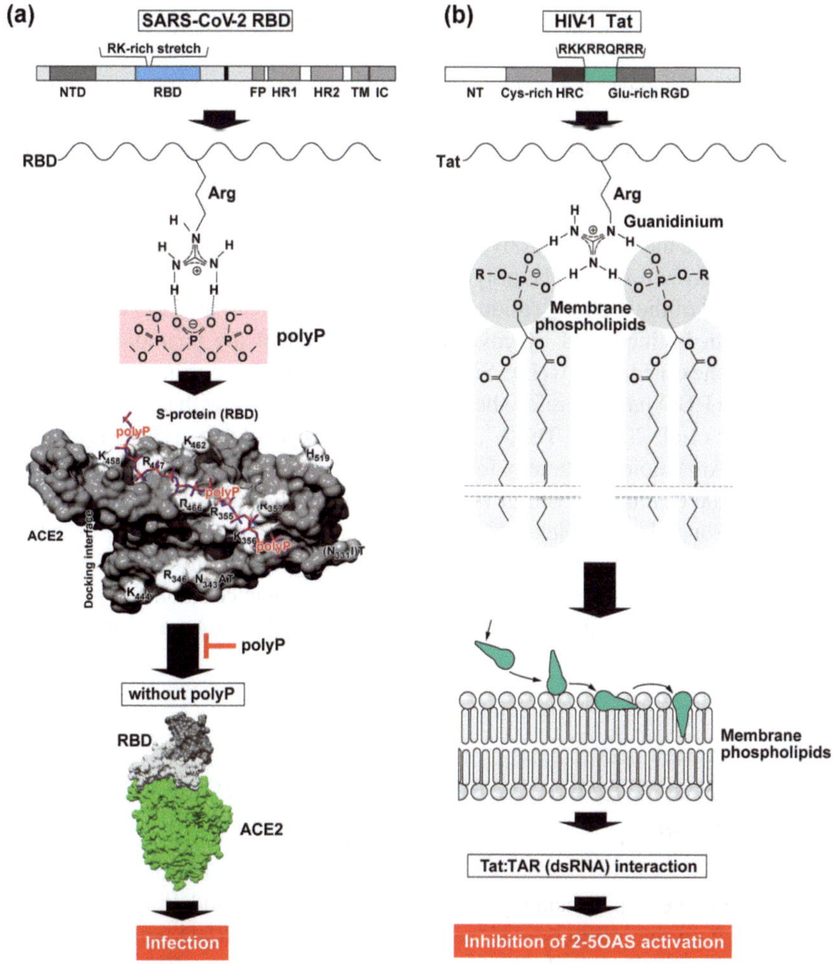

Fig. 7.3 Electrostatic interactions of the guanidinium group of the arginine-rich regions within the RBD of the SARS-CoV-2 spike protein and the HIV-1 Tat protein with phosphate. Below the schemes of the proteins, the hydrogen bond formation and electrostatic interaction of the arginine guanidium groups of (**a**) the SARS-CoV-2 with its RK-rich stretch and (**b**) the Tat protein with its basic RKKRRQRRR segment (**a**) with an acid anhydride bound phosphate within the polyP polymer and (**b**) with the ester-bound phosphate of two adjacent phospholipid molecules at the outer cell membrane surface are shown. (**a**) PolyP interacts with the guanidium groups of multiple basic amino acids (arginine and lysine) of the SARS-CoV-2 RBD, resulting in inhibition of RBD-ACE2 attachment and cell infection. (**b**) The binding of Tat via its basic amino acid segment to the membrane lipids enables the passage of this protein through the cell membrane. Within the cell, Tat can bind to the HIV-1 TAR stem-loop and thereby prevent binding and activation of the 2′,5′-oligoadenylate synthetase (2-5OAS) involved in the cellular innate immune defense against virus

formation of fusion pore by inducing membrane multilamellarity (Allolio et al. 2018).

The RBD sequence is highly conserved among the pathogenic human coronaviruses. Interestingly, based on modelling experiments—and this might be important with regard to a potential clinical application of polyP—the asparagine/tyrosine exchange at amino acid position 501 (N501Y mutation) of the RBD sequence of the British SARS-CoV-2 variant B.1.1.7 (Rambaut et al. 2020) is not located at or close the cationic groove (Müller et al. 2020a). Therefore, it is unlikely that this mutation will impair the polyP:RBD interaction.

The potential of the arginine residues on a protein surface to undergo electrostatic interactions can be increased by covalent modification of the guanidinium group with 1,2-cyclohexanedione (CHD; Suckau et al. 1992). The surface-exposed Arg residues of the RBD have therefore been modified by reaction with CHD in a borate buffer (Müller et al. 2020a). The N^7,N^8-(1,2-dihydroxycyclohex-1,2-ylene)-L-arginine (DHCH-Arg) moieties thereby formed protrude through the hydrate shell of the protein molecule (Patthy and Smith 1975; Domingo et al. 2017). These DHCH-Arg residues are strong electrophiles. Their reactivity is even increased by the formation of a dynamic gradient of water dipole molecules around the modified Arg that stick out into hydration shell of the protein. In contrast, the resonance stabilized positively charged guanidinium group of the non-modified Arg only shows a comparatively low reactivity.

7.4.7 *PolyP as Effective Inhibitor of the Spike:ACE2 Interaction*

In order to validate experimentally the prediction made in modelling studies, the effect of polyP on the interaction of the SARS-CoV-2 spike RBD with the cellular receptor protein ACE2 was determined using a binding assay. This ELISA-like assay consisted of the recombinant ACE2 bound to the solid phase and a biotin-labeled recombinant RBD present in the fluid phase (Shang et al. 2020). PolyP, the predicted inhibitor of the RBD:ACE2 interaction, was preincubated with the ACE2 prior to the addition of the recombinant protein to the well plate. The quantity of RBD bound to the receptor was determined with labeled streptavidin using a chemiluminescence-based detection method. The results revealed that polyP significantly inhibited the RBD:ACE2 binding (Neufurth et al. 2020; Müller et al. 2020a). Already in the presence of 1 μg mL^{-1} of polyP, a reduction of binding to 72% was measured (Fig. 7.4a). The inhibition increases at a higher concentration of 30 μg mL^{-1} of polyP, which reduced the binding between RBD and ACE2 to 42%. In this experiment polyP with a chain length of 40 P_i units (Na-polyP$_{40}$) has been used. Unexpectedly, a reduction of the chain length of the polymer did not abolish the inhibitory activity of polyP. The very short Na-polyP$_3$, with only 3 P_i residues, turned out to be similarly active in inhibiting the binding propensity of RBD to the

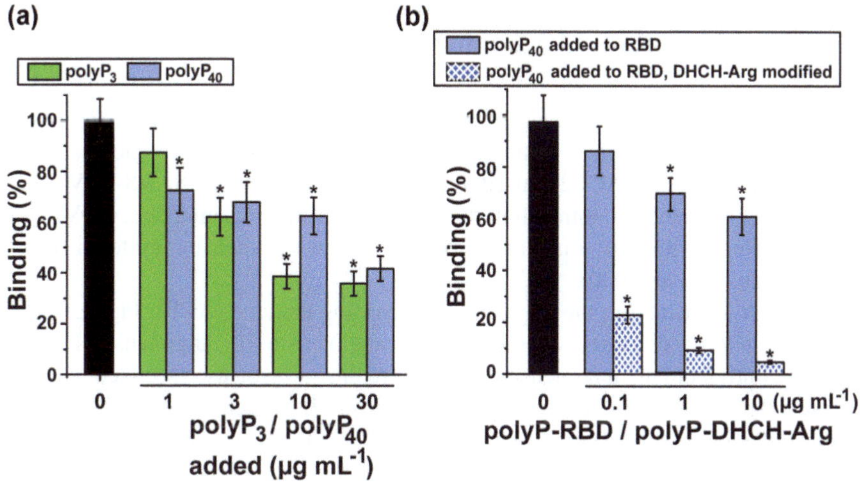

Fig. 7.4 Inhibition of RBD:ACE2 interaction by polyP. (**a**) Effect of increasing concentrations of Na-polyP$_{40}$ and Na-polyP$_3$ on the ability of the RBD to bind to ACE2. (**b**) Increase of the inhibitory activity of Na-polyP$_{40}$ after modification of the Arg residues in the RBD with CHD. The binding strength in the absence of polyP was set at 100%. Adapted with permission from Müller et al. (2021b). Copyright 2021, Elsevier

ACE2 receptor (Neufurth et al. 2020; Müller et al. 2020a); Fig. 7.4a. A possible explanation could be that inhibition is saturable because of a steric hindrance of the polyP$_{40}$ molecules at the Arg residues. Alternatively, the inhibitory action of polyP is due to the interaction only with a short segment of the cationic groove, which is short enough to be covered by the short-chain polyP$_3$ oligomers.

The latter assumption has been confirmed experimentally by chemical modification of the Arg residues at the RBD surface with CHD. The treatment of the Arg residues with CHD increases the electron density distribution around the guanidinium groups of the Arg residues at the RBD surface, allowing a stronger electrostatic interaction with the polyanionic polyP chains. Indeed it was found that the inhibitory potency of polyP in the RBD:ACE2 binding assay markedly increases (Müller et al. 2020a).

A significant 77% reduction of the binding efficiency of the resulting DHCH-Arg to the RBD surface was measured at a Na-polyP$_{40}$ concentration of 0.1 µg mL^{-1}, compared to the viral binding protein with non-modified Arg, which caused a reduction of binding strength by only to 12% (Fig. 7.4b). At a higher concentration of 10 µg mL^{-1}, Na-polyP even caused a reduction by 97% of the ACE2 binding propensity of the DHCH-Arg-modified RBD, while only an inhibition of binding by 40% was measured for the non-modified RBD.

It should be noted, the level of polyP in blood, with a chain length of 40 P$_i$ units, is between 0.5 and 3 µg mL^{-1} and therefore sufficient to block the RBD:ACE2 interaction.

7.5 The Consequence of the Spike:ACE2 Interaction: Impairment of ACE2 Function

ACE2 is a carboxypeptidase that splits off the *C*-terminal phenylalanine (Phe) from angiotensin II (consisting of 8 amino acids) under formation of angiotensin 1–7 (7 amino acids). Since ACE2 is used as the host cell receptor by SARS-CoV-2 and becomes internalized and intracellularly degraded during virus infection, COVID-19 is associated with an imbalance of the renin-angiotensin system (RAS) (Vaduganathan et al. 2020).

ACE2 is distinct from the angiotensin converting enzyme (ACE); Fig. 7.5. This metalloproteinase converts angiotensin I to angiotensin II. Angiotensin I is the cleavage product of angiotensinogen by renin, a protease secreted by the juxtaglomerular cells in the kidneys. Angiotensin II, which is formed by ACE, binds to the angiotensin II type 1 receptor (AT_1R) and acts as a potent vasoconstrictor. It also causes inflammation, thrombosis, and fibrosis, and activates the production of reactive oxygen species (ROS).

ACE2 counteracts these physiological effects caused by ACE. Angiotensin 1–7 which is formed by ACE2 from the ACE product angiotensin II, binds to a receptor, which is different from AT_1R. Binding of angiotensin 1–7 to this receptor, the Mas receptor (MasR), induces the opposite effects to those found by activation of AT_1R. Activation of MasR results in vasodilation, anti-inflammation, anti-thrombosis, anti-fibrosis, and elimination of ROS activity. Therefore, if after infection of cells with SARS-CoV-2 the ACE2 receptor becomes downregulated, the activity of AT_1R cannot be counterbalanced and an uncontrolled, excessive production of pro-inflammatory cytokines ("cytokine storm") characteristic for severe cases of COVID-19 occurs (Zhang et al. 2015).

Interestingly, activation of the RAS pathway is found in many pathological conditions that are associated with an increased risk of fatal outcomes of COVID-19 disease, such as diabetes, hypertension, obesity, cardiovascular diseases, and cancer (Verdecchia et al. 2020; Tseng et al. 2020). Also, the age-dependent decrease in ACE2 expression could be a decisive factor for the poorer prognosis in COVID-19 patients with increasing age.

7.6 Virus Entry

Following the proteolytic cleavage at the S1/S2 boundary, the released S1-subunit (*N*-terminal part of the spike S-protein with the RBD) is retained on the cellular ACE2 receptor. On the other hand, the S2-subunit (*C*-terminal part) with its transmembrane domain remains bound to the envelope of the viral particles (Tang et al. 2020). This subunit contains the fusion peptide as well as the HR1 and HR2 heptad-repeat segments. These two segments consist of a repetitive heptapeptide sequence, HPPHCPC, composed of hydrophobic (H), polar or hydrophilic (P) and charged

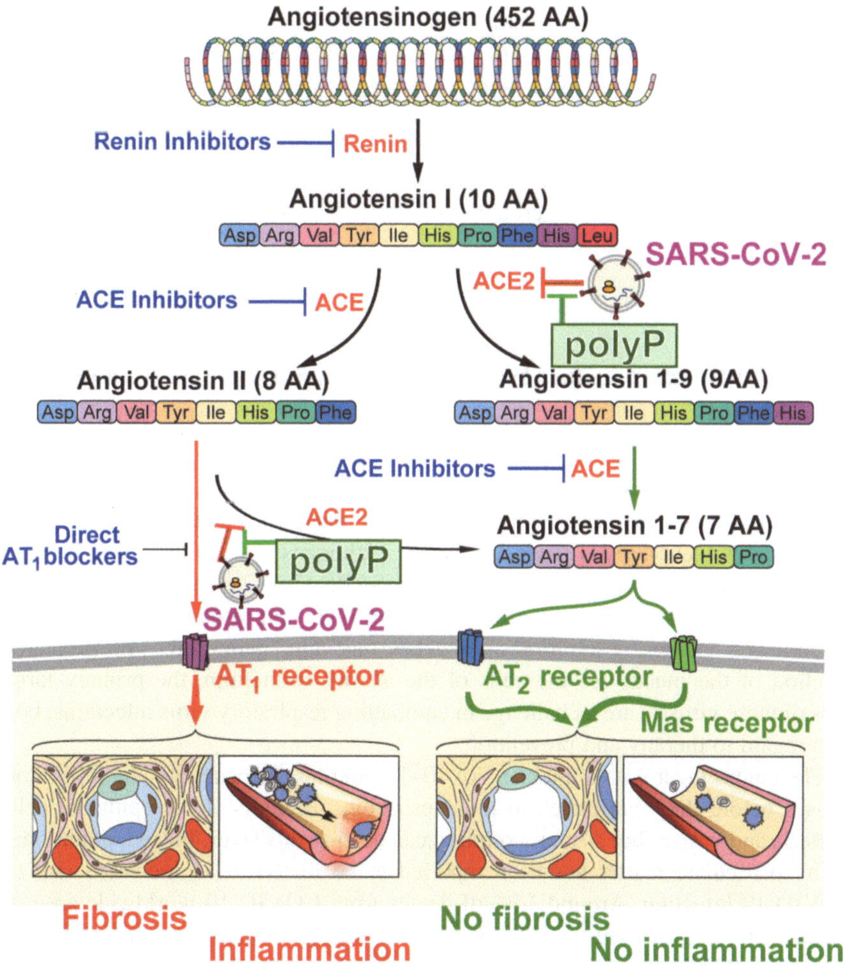

Fig. 7.5 Imbalance of the opposing effects of the ACE—angiotensin II—AT1 receptor and ACE2—angiotensin 1–7—Mas receptor pathways due to the binding of SARS-COV-2 to its host cell ACE2 receptor. Angiotensinogen is cleaved by the protease renin to angiotensin I, which is converted either by ACE to angiotensin II or by ACE2 to angiotensin 1–9. Angiotensin II is cleaved to angiotensin 1–7 by ACE2. The latter product is also formed from angiotensin 1–9 by ACE. The attachment of SARS-CoV-2 to ACE2, which serves as the virus receptor, impairs the function (or downregulates the expression) of ACE2, which leads to an imbalance in both signaling pathways. While angiotensin II activates the AT1 receptor and causes inflammation, in addition to vasoconstriction, thrombosis, fibrosis, and ROS production, the binding of angiotensin 1–7 to the Mas receptor prevents these deleterious effects

(C) amino acid residues. In the next step, a pre-hairpin is formed, whereby the S2 subunit is linked to the plasma membrane through its HR1 domain and to the viral envelope through its HR2 region (Kawase et al. 2019; Shang et al. 2020; Tang et al. 2020). The fusion process is initiated by a conformational change of the hydrophobic fusion protein within the S2 subunit, which then becomes inserted into the host cell membrane (pre-hairpin formation) (Harrison 2015). Thereby, both membranes approach each other, with HR1 being closer to the cell membrane and HR2 being closer to the virus membrane. Then, a six-helix bundle structure is formed by backfolding of HR2 to HR1 (Simmons et al. 2013; Xia et al. 2020). As a result, the virus membrane is pulled in the direction of the host cell membrane, so that the two membranes can fuse (Eckert and Kim 2001; Xia et al. 2018). This enables the virus RNA to enter the host cell.

7.7 The Barrier to be Overcome by the Virus: The Mucus Shield

The mucus of the respiratory tract but also of the gastrointestinal tract has an important function as a barrier that shields the underlying epithelial layer against attack and infection by viruses and other pathogens. The mucus also plays a prominent role in the initial defense against SARS-CoV-2 infection and infection by other respiratory viruses, like influenza. This layer can be damaged/disrupted by bacteria and viruses. Therefore, approaches that aim to reinforce the protective function of the mucus for the cells of the airway epithelium, the primary target cells of these viruses, are a challenge in combatting respiratory virus infections, both with regard to therapy and prevention.

The mucus layer with a thickness of 10–15 μm must be penetrated by respiratory viruses before they can attach to their respective receptors at the epithelial cells. Bacteria and viruses but also air pollution can destroy this layer. Indeed, pollution of air by particulate matter has been reported to be linked to an increased risk for COVID-19 infection. Around 15% of deaths from COVID-19 worldwide are estimated to be due to long-term exposure to PM2.5 (fine particles with a diameter $\leq$2.5 μm) (Pozzer et al. 2020). The adverse effect is caused by increased damage to the cells of the airways, especially to the ACE2 expressing respiratory cells (Aztatzi-Aguilar et al. 2015; reviewed in Paital and Agrawal 2021). At present, drugs that restore the physiological composition and functional organization of the mucus are not yet known. Most drugs available acting on the mucus level are neuraminidase inhibitors of influenza A virus, which are not applicable to other respiratory viruses (Ison and Hayden 2017). Apart from neuraminidase inhibitors, known compounds acting on the mucus level primarily only improve the clearance mechanism, either by reducing the viscosity of the mucus (mucolytics, e.g., by breaking disulfide bonds like *N*-acetylcysteine) or increasing its hydration/volume (expectorants) or transport (mucokinetics) or suppressing hypersecretion (mucoregulators).

The mucus is composed of a dense two-layer network of hierarchically organized mucin molecules. These glycoproteins have a high molecular mass ranging from 10^4 to 10^6 Da (Guzman-Aranguez and Argüeso 2010). Their glycan components can amount up to 90% of this mass. Two different types of mucins are distinguished; the secreted mucins which comprise both oligomeric gel-forming mucins, such as MUC5AC, and non-polymeric mucins, and the tethered, cell membrane-associated (membrane-spanning) mucins, such as MUC1 (Thornton et al. 2008; Petrou and Crouzier 2018). They are secreted by the goblet cells of the respiratory tract. The gel-forming mucins constitute the upper layer of the mucus, the epithelial lining fluid, which is located above the periciliary layer that is built by the membrane-associated mucins that surround the epithelial cell layer.

In addition to the goblet cells, the respiratory epithelia are composed of the ciliated cells which are involved in mucociliary transport (velocity of 4–10 mm/min; Hansson 2019), as well as the nonciliated cells and the basal cells.

7.7.1 Crossing the Mucus *Barrier*

The mucus forms a fine-meshed network with spacings between 20 and 1800 nm, which allows the diffusion of small molecules, including virus-like particles (Olmsted et al. 2001). Influenza A viruses with a 80–120 nm can penetrate through the mucus, facilitated by their sialic acid cleaving neuraminidase which reduces ionic repulsions (Kaler et al. 2020). The SARS-CoV-2 virus (size range from 60 to 140 nm; Bar-On et al. 2020) is assumed to be able to penetrate through the mucus layer, based on the results of experiments with similar-sized nanolipoparticles (Lai et al. 2009).

7.7.2 Interaction of PolyP with Mucin

The respiratory epithelium is the primary site of infection with SARS-CoV-2. Therefore, if applied for prevention or therapy of COVID-19, it is reasonable to apply polyP in the form of a spray or as a mouth rinse, either as soluble polyP (sodium salt) or as polyP nano/microparticles (Ca^{2+} or Mg^{2+} salt). The question arises whether anionic mucins can interfere with the antiviral action of polyP, to block the binding of the viral RBD to the cellular ACE2 receptor. In order to clarify this question, combinatory experiments with polyP and mucin have been performed. The results revealed that polyP is still fully active in a mucin-containing matrix (Müller et al. 2020b). At a mucin concentration of 1 mg mL^{-1} polyP was even found to inhibit the RBD:ACE2 interaction at a concentration lower than that of polyP found in blood. At a polyP concentration of 10 μg mL^{-1} and 100 μg mL^{-1} addition of mucin did not reduce the polyP-caused inhibition of receptor binding of the SARS-CoV-2 spike protein as measured in the RBD:ACE2 binding assay.

Even more important, studies on the effect of polyP on mucin gene expression with A549 cells (human alveolar basal epithelial cells) revealed that polyP also exhibits morphogenetic activity in this cell system (Wang et al. 2018; Müller et al. 2020b). These cells express the ACE2 receptor. The steady-state expression of both

MUC1 and *MUC5AC* gene in A549 cells was found to be significantly upregulated after exposure of the cells to polyP. In addition, it was shown that incubation of cultures of the cells with polyP leads to a rise of production of extracellular ATP mediated by the cell membrane-associated ALP and ADK (Müller et al. 2020b). These results confirmed that polyP acts protectively also in the alveolar cell system.

In addition, polyP might affect mucin production by a second, different mechanism: through facilitating the "unpacking" of mucin (Kesimer et al. 2010) during the release of the molecules from goblet cells. It has been shown that mucin synthesis is associated with a reversible condensation and expansion process of the mucin molecules that depends on pH and ions (Ridley and Thornton 2018). Intracellularly, mucin, after glycosylation and intermolecular disulfide linkage, is stored, together with Ca^{2+} ions, in a condensed dehydrated state in granules with an acidic pH (pH 5.5). The Ca^{2+} ions shield the negative charges of the molecule, which are due to the sialic acid and sulfate residues at the glycan side chains. During release of the granules, an explosive expansion of the condensed material occurs, which is paralleled by removal of the Ca^{2+} ions and their replacement by Na^{+}. This process results in an increased hydration of the polymer. Due to its ability to complex Ca^{2+} ions, it seems to be likely that polyP also promotes mucin formation during the release step of the polymer.

7.7.3 Antimicrobial Proteins and Surfactant Proteins

The barrier function of the mucus is supported by a series of antimicrobial proteins that contribute to the antimicrobial defense in the respiratory tract, as well as surfactant proteins, which are present in the mucus. Antimicrobial proteins, such as lysozyme, β-defensin, and cathelicidin (LL-37), are also antivirally active and can prevent a secondary bacterial infection through the mucus layer (Ganz 2002; Ahmed et al. 2019). Surfactant proteins, such as SP-A and SP-D, belong to the collectin family of proteins (Wright 2005). The trimeric polypeptide chains of these proteins, which assemble into oligomers, are characterized by a collagen-like *N*-terminal region and *C*-type (calcium dependent) carbohydrate-recognition domain (CRD), or lectin domain. SP-A and SP-D bind via their lectin domains and thereby neutralize a larger number of respiratory bacteria and viruses via agglutination and opsonization (Whitsett and Alenghat 2015).

7.7.4 Energy Requirement of the Mucus Layer: Strengthening Through PolyP

The synthesis as well as the maintenance of the functionally active state of the mucus, including the synthesis of its components and its composition and organization, requires metabolic energy in the form of ATP. Inhalation of ATP was found to promote the release of mucin from the goblet cells (Shin et al. 2000). It has also been reported that ATP induces the release of MUC5AC from human bronchial preparations in vitro (Roger et al. 2000). This effect caused by extracellular ATP most likely occurs via the $P2Y_2$ receptor on the airway epithelium (Shishikura et al. 2016). Also, the mucociliary clearance mechanism that removes inhaled pathogens such as viruses from the airways system, needs energy. This mechanism, driven by the ciliary beating, depends on the presence of extracellular ATP that might be released via an autocrine mechanism (Droguett et al. 2017). It has been proposed that the release of ATP from airway epithelial cells occurs via pannexin channels, such as Panx1 (D'hondt et al. 2011).

Infection of cells with virus can impair the energy state of a cell. The amount of ATP released from cells, even from uninfected cells, is rather small. Most likely the low extracellular ATP level is further decreased after virus infection. This lack of energy could be compensated by polyP, which acts as the ATP generator in the extracellular environment. The ALP catalyzes the phospho-transfer from polyP to ADP that can be further converted to ATP by ADK, which is present in the airway mucus (Picher et al. 2003).

7.7.5 PolyP and Extracellular Chaperone Activity

Clusterin is an extracellular chaperone-like protein that is also present in the mucus (Sol et al. 2016). This protein most likely plays a role in the control of the correct folding and assembly of extracellular proteins that are damaged by stress or under pathological conditions. In addition, it can induce the expression of MUC5AC in airway epithelial cells (Bae et al. 2018). Clusterin contains an ADP binding motif (GxGxxG and GxxGxG; Tsuruta et al. 1990). It is assumed that extracellular aberrant proteins bound to clusterin are taken up by cells after binding to a cell surface heparan sulfate receptor and then degraded intracellularly (Itakura et al. 2020); Fig. 7.6. Based on studies of the effect of polyP on cells exposed to the Alzheimer peptide Aβ25-35, it has been proposed that binding of extracellular ADP, which is produced from polyP, to the ADP/ATP binding motif of clusterin can prevent misfolding of the β-amyloid precursor protein (Müller et al. 2017b). It is likely that clusterin is involved in the maintenance of the integrity of mucin network (Schepler et al. 2021).

The expression of clusterin is induced by stress and apoptosis. In SARS-CoV-2 infection, apoptosis is induced by the SARS-CoV-2 ORF3a protein, which activates caspase-8 (Ren et al. 2020), resulting in cleavage of Bid, a Bcl2 interacting protein, and reactive oxygen species (ROS) production (Fig. 7.6).

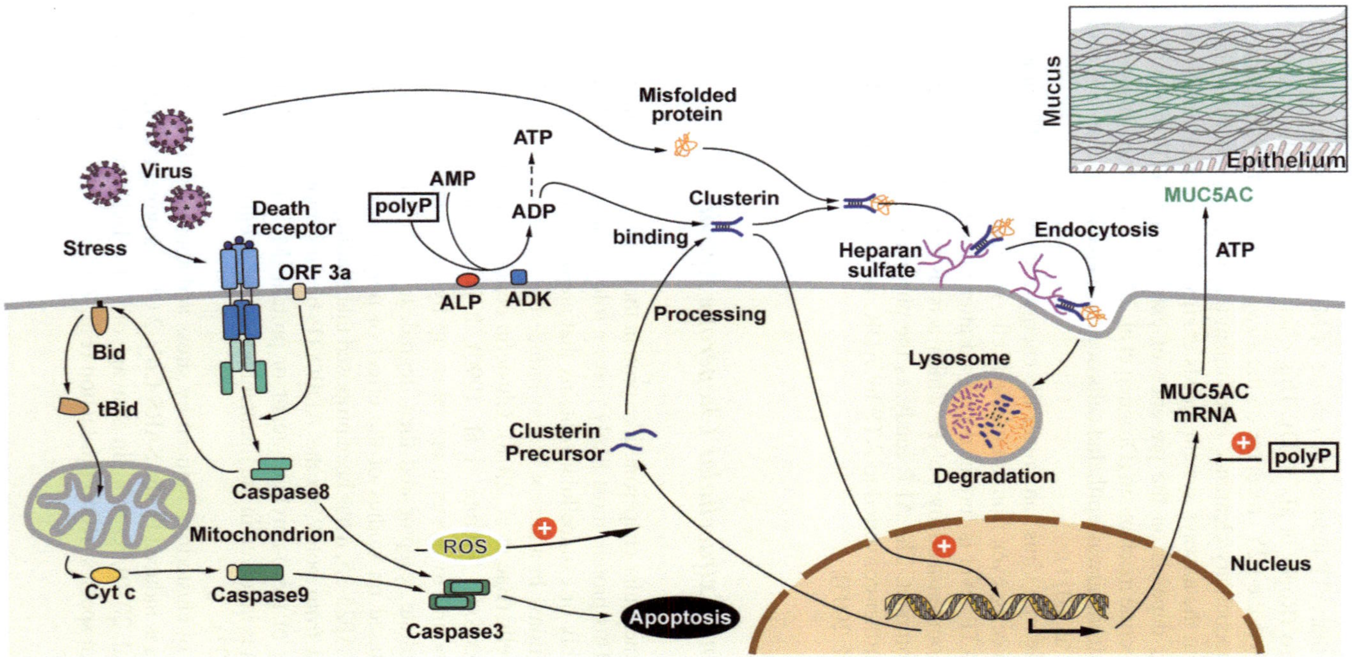

Fig. 7.6 Effect of polyP on ADP/ATP-dependent processes involved in the formation and maintaining of function of the antivirally protective airways mucus layer. PolyP is functioning as a source of ADP/ATP, which is generated via the combined ALP/ADK action on the outer cell membrane surface. ADP binds to the extracellular chaperone clusterin, which is induced as a result of virus infection (or stress and apoptosis). The ADP-clusterin recognizes misfolded proteins, which are subjected to lysosomal degradation by endocytic uptake of the clusterin-misfolded protein complexes after binding to heparan sulfate on the cell surface of the respiratory epithelium. In addition, polyP (as well as clusterin) promotes the ATP-dependent formation, release, and assembly of mucins, like MUC5AC, in the mucus layer. On the left side, the induction of apoptosis by the SARS-CoV-2 ORF3a protein is shown, which involves the activation of caspase-8 and cleavage of Bid to truncated tBid, which translocates to mitochondria to trigger cytochrome c release and ROS production, resulting in an increased clusterin formation

7.8 PolyP and Cytokine Expression

7.8.1 Interaction with the Interferon System

The interferon system plays a prominent role in the innate immune defense against viral infection. SARS-CoV-2 induces the host cell to express primarily type I interferons (mainly interferon-α and interferon-β), in addition to type III interferon (interferon-λ). The expression of type I interferons, like that of other cytokines, is induced via recognition of pathogen-associated molecular patterns (PAMPs; such as viral double-stranded RNA, dsRNA) and damage/danger-associated molecular patterns (DAMPs, which are released from damaged cells after virus infection) by pattern recognition receptors (PRRs), such as Toll-like receptors (TLRs). Interferon induction then leads to the expression of a number of antivirally acting interferon-stimulated genes (ISGs) via activation of the transcription factors STAT1 and STAT2.

Besides activating interferon production, recognition of viral dsRNA by PRRs results in the activation of antiviral pathways within the host cells, the most important are the 2′,5′-oligoadenylate synthetase (2-5OAS)/ribonuclease L (RNase L) and protein kinase R (PKR) pathways (Schröder et al. 1994b).

SARS-CoV-2 infection of cells leads to an impairment of type I interferon production (Matsuyama et al. 2020). It has been proposed that this effect is caused by the viral NSP1 and ORF6 proteins, which interfere with the STAT1 and STAT3 pathways (Matsuyama et al. 2020). Only a weak induction of interferon has also been detected in A549 cells, which we used in our studies (Müller et al. 2020b), after infection with SARS-CoV-2 (Li et al. 2021). In contrast, SARS-CoV-2 was found to strongly activate the OAS and PKR in these cells (Li et al. 2021).

It is striking that type I interferons also induce the expression of ACE2, just that protein that is used by SARS-CoV-2 as the receptor (Garcia-del-Barco et al. 2021). This finding might be contradictory to a potential application of interferon in treatment of COVID-19 patients. On the other hand, SARS-CoV-2 infection is known to lead to an imbalance of the renin-angiotensin system, through cellular uptake and degradation of the receptor (see above), which, consequently, could be mitigated by interferon. Therefore, a therapeutic application of interferon for COVID-19 patients, at least in the early stage of the disease, seems to be reasonable (Garcia-del-Barco et al. 2021).

7.8.2 Induction of the Antiviral 2-5A Pathway and Escape of the Virus

The interferon-induced OAS pathway is an important antiviral mechanism involved in the establishment of an antiviral state of cells (Schröder et al. 1989, 1990a, 1992). Binding of viral dsRNA to OAS results in activation of the enzyme, OAS, which forms 2′,5′-linked oligoadenylates (2-5A) from ATP (Fig. 7.7). 2-5A, in turn, activates RNase L, which after dimerization, degrades viral but also cellular single-stranded RNA. The dsRNAs which activate OAS also comprise RNA stem-loop structures that are present in viral mRNAs, such as the TAR sequence of HIV-1 (Schröder et al. 1990b). The 2-5A is inactivated by a poly(A)-specific exoribonuclease (Schröder et al. 1980), which degrades not only the poly (A) sequence of mRNA (3′,5′-linked) but also 2′,5′-oligoadenylate (Müller et al. 1980). This 2-5A system, which is impaired with age (Kuusksalu et al. 1995), is also active in SARS-CoV-2 infected cells (Sa Ribero et al. 2020).

However, SARS-CoV-2 has developed mechanisms to escape from this antiviral pathway. These mechanisms consist not only of the reduction of interferon expression, but might also involve the degradation of 2-5A by the viral NS2 protein, which might act as a phosphodiesterase, as shown for other members of the beta-coronavirus family (Zhang et al. 2013); Fig. 7.7.

It has been reported that polyP suppresses the expression of interferon-β and interferon-regulated genes (Roewe et al. 2020). However, these studies have been performed with long-chain polyP produced by bacteria (300–1000 P_i residues) (Roewe et al. 2020) or polyP with medium-chain lengths ($polyP_{75}$; Chrysanthopoulou et al. 2017; Suess et al. 2019). The effect of shorter polyP chains is not known. It has been suggested that the effects of short-chain polyP on endothelial cells occur via the purinergic $P2Y_1$ and RAGE receptors (Holmström et al. 2013; Dinarvand et al. 2014).

The risk of a severe course of COVID-19 as well as the susceptibility to SARS-CoV-2 infection is correlated with high plasma levels of the OAS1 isoform of OAS before infection (Zhou et al. 2021). Interestingly, just this isoform (p46 isoform), which is protective against COVID-19, has been shown to be introduced by Neanderthals into modern human population in Europe about 50,000–60,000 years ago (Zeberg and Pääbo 2021). This isoform is found also extracellularly (Kristiansen et al. 2010). In modern humans several isoforms of OAS1 mostly exist that are less active and are generated by altered splicing of the OAS1 transcript while the Neanderthal p46 isoform exhibits a high enzyme activity.

The effect of polyP on the second enzyme that is activated by dsRNA and SARS-CoV-2 infection, the PKR, has not yet been studied. This enzyme, after binding of dsRNA, becomes activated by autophosphorylation and phosphorylates the eukaryotic initiation factor eIF2α, resulting in an inhibition of translation of viral proteins.

7.8.3 PolyP: Mitigating the "Cytokine Storm"?

The RNA of SARS-CoV-2 can be recognized by cellular PRRs, e.g., Toll-like receptors (TLRs) which are located in the endosomal membrane. After activation by viral dsRNA these receptors initiate an intracellular signaling cascade which is part of the innate immune response of cells against virus infection (Totura et al. 2015; Fitzgerald and Kagan 2020). The activation of TLR3 with dsRNA leads,

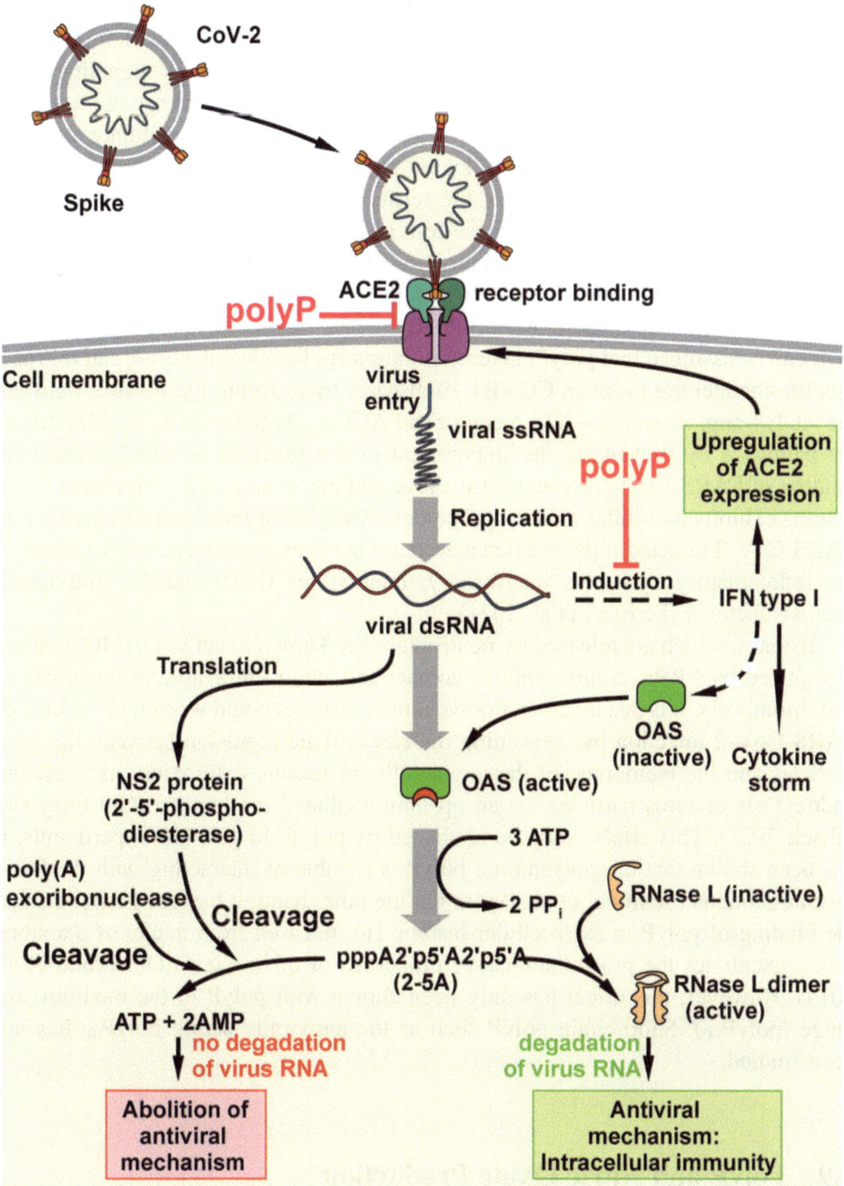

Fig. 7.7 Interaction of SARS-CoV-2 with the interferon-inducible 2-5A pathway. Interferon (IFN) induces the expression of OAS, which after binding of dsRNA activates RNAse L, resulting in degradation of viral RNA ("intracellular immunity"). 2-5A might be inactivated by viral NS2 protein, which acts like the 2-5A-degrading poly(A)-exoribonuclease. The latter enzyme, which is involved in catabolism of the poly(A) tail of mRNA, has been shown also to cleave 2′,5′-phosphodiester bonds in addition to 3′,5′-phosphodiester linkages

among other things, to induction of expression and release of type I interferons (Totura et al. 2015; Kawai and Akira 2007). Besides the release of interferons, mediated through the interferon regulatory factor 3 and the NF-κB signaling pathway, a series of cytokines involved in the adaptive immune response are secreted, including the interleukins-1, -6, and -18 and the pro-inflammatory tumor necrosis factor α (Vabret et al. 2020). In parallel with this process, an activation of T helper lymphocytes can occur resulting in the release of interferon-γ, interleukin-6, and other cytokines or chemokines by these cells (Patra et al. 2020; Tay et al. 2020). The exaggerated release of these cytokines in COVID-19 patients, in particular of interleukin-6, termed "cytokine storm" (Melkamu et al. 2013), is associated with a poor prognosis of the disease (Magro 2020).

It can be assumed that polyP, at least in the short-chain length range, can mitigate this life-threatening event in COVID-19 disease, by restoring the balance between the ACE—angiotensin II—AT1 receptor and ACE2—angiotensin 1–7—Mas receptor pathways by preventing the impairment of the function of ACE2 caused by binding of SARS-CoV-2. Recent results revealed that even polyP with longer chain lengths exhibits anti-inflammatory properties. Analysis of the effect of $polyP_{120}$ on SARS-CoV-2 infected cells revealed a decrease in the expression, at mRNA level, of the inflammatory cytokines interferon-γ, interleukins 6, 10 and 12, and tumor necrosis factor-α (Ferrucci et al. 2021).

Histones, which are released by neutrophils, are known to act as DAMPs that are recognized by PRRs, contributing to contact activation, intravascular coagulation, and thrombosis. It is discussed that polycationic histones could worsen the course of SARS-CoV-2 infection by weakening the electrostatic repulsion between the viral particles and the membrane of the target cells, increasing cytotoxicity or inducing endocytosis of virus particles via an opsonin-mediated mechanism (Ginsburg and Fibach 2021). This effect could be abolished by polyP. In gel-shift experiments, it has been shown that the polyanionic polyP is capable of interacting with the basic histone proteins (Schröder et al. 1999). On the other hand, it has been reported that the binding of polyP to extracellular histone H4, released from nuclei of damaged cells, potentiates the pro-inflammatory properties of this protein (Dinarvand et al. 2014). However, this effect has only been shown with polyP in the medium-size range ($polyP_{70}$). Short-chain polyP such as the antivirally acting $polyP_{40}$ has not been studied.

7.9 PolyP and Nitric Oxide Production

The effects of polyanionic polymers like polyP on some aspects of the innate immune system deserve further attention. These include the potential effects of the polymers on nitric oxide (NO) production. It has been reported that NO inhibits the replication of SARS-CoV (Akerström et al. 2005), most likely on the level of viral protein and RNA synthesis (Akerström et al. 2009). NO also possesses anti-inflammatory activity and plays a role in pulmonary vascular function and has been proposed to present a possible target for therapy of COVID-19 (Adusumilli et al. 2020). The inducible nitric oxide synthase (iNOS) is increased during infection by SARS-CoV-1. It has been reported that polyP suppresses the expression of iNOS, but this effect (inhibition versus stimulation) seems again to depend on the chain length of the polymer (Harada et al. 2013; Roewe et al. 2020) and needs to be clarified.

7.10 PolyP and Platelet Activation: Prothrombotic Versus Antithrombotic Effects

The effect of polyP on clotting has been thoroughly investigated by several groups. These studies revealed that polyP can accelerate the clotting process via different mechanisms, involving binding to factor XII/XIIa, factor V activation, and fibrin clot formation (Müller et al. 2009; Morrissey et al. 2012; Baker et al. 2018). However, the results of these studies, which also indicate a pro-inflammatory effect of polyP (Müller et al. 2009), must be relativized by the fact that the effects of polyP strongly depend on the chain length of the polymer and are not or only marginally found with polyP in the physiological size range (Smith et al. 2010; Choi et al. 2011). It rather seems that physiological polyP has an inhibitory effect on the clotting process (Yang et al. 2017). These results are also confirmed by studies based on clot waveform analysis (Kondo et al. 2019; Wakui et al. 2019).

Activation of platelets as a result of polyP-caused factor XII activation can induce the release of nuclear DNA-based extracellular fiber networks by neutrophils (neutrophil extracellular traps, NETs) that can trap pathogens, e.g., virus, but also platelets, thereby contributing to hypercoagulability and thrombosis. PolyP has been reported to act as a NET inducer, but again, this effect has only been observed for long-chain (bacterial) polyP (Raghunathan et al. 2019) or medium-sized polyP ($polyP_{75}$; Chrysanthopoulou et al. 2017).

A severe complication of therapy with the organic polyanion heparin is the development of heparin-induced thrombocytopenia (HIT) (Arepally 2017). HIT is caused by antibodies against complexes of heparin and platelet factor 4 (PF4). These antibodies bind to the FcγRIIA receptors on platelets, resulting in platelet activation, thrombin generation, and hypercoagulability, leading to thrombosis. It has been reported that PF4 also forms complexes with polyP that could be recognized by the antibodies (Cines et al. 2016). However, the observed effects again show a strong dependence on the chain length of the polymers and significant effects in the physiological concentration range have only been found with polyP sizes of ≥ 60 P_i units. Moreover, it has been published that low-chain polyP rather exhibits antithrombotic activity (Church et al. 1988; Ozaki et al. 2021). Investigations on the effect of polyP and other phosphate-containing polyanions on the glycosaminoglycan-binding plasma proteinase inhibitors antithrombin III (ATIII) and heparin cofactor II (HCII), which mediate the antithrombin activity of heparin, revealed that polyP with chain lengths from 4 to 65 P_i units does not affect the ATIII-thrombin reaction (Church et al. 1988). It accelerated the HCII-thrombin reaction. Based on these results it was concluded that the antithrombotic effect of polyP and other phosphate-containing polyanions is mediated by activation of HCII but not through ATIII. The apparently positive effect of polyP in preventing thrombosis, a major problem in severe cases of COVID-19, will be of interest especially for a potential therapeutic application of polyP; further studies are needed.

7.11 Antibacterial Effects of PolyP

In addition to antiviral activity, polyP exhibits antibacterial activity. The antibacterial effects of polyP might also be important with regard to virus infection, because bacteria could destroy the integrity of the mucus layer of the respiratory tract, facilitating infections with respiratory virus, like SARS-CoV-2. PolyP has a strong antibacterial activity against Gram-positive bacteria, while Gram-negative bacteria are more resistant to the polymer (Obritsch et al. 2008; Moon et al. 2011; Lorencová et al. 2012; Müller et al. 2017a). This effect can be partially explained by oxidative stress caused by polyP (Moon et al. 2011). The effect of polyP on Gram-positive bacteria has also been attributed to damage to the bacterial cell envelope causing leakage of the cells (Lee et al. 1994). Most likely this mechanism is caused by the chelation of essential metal ions by polyP that competes with metal ion binding teichoic acid chains in the cell walls of Gram-positive (but not of Gram-negative) bacteria (Lee et al. 1994).

7.12 Potential Therapeutic Application of PolyP for Treatment of SARS-CoV-2 Infection

Three phases can be distinguished during progression of COVID-19 disease: pulmonary phase, pro-inflammatory phase, and prothrombotic phase (Lee and Choi 2021); Fig. 7.8. The pulmonary phase describes the infection of the host cells with SARS-CoV-2 and is characterized by the occurrence of interstitial pneumonia and acute respiratory distress syndrome (ARDS). The ARDS is caused by an impairment of the function of ACE2, which is used by the virus to attach to the host cell. As a consequence, an imbalance of the renin-angiotensin system is observed. A characteristic feature of the pro-inflammatory phase is the overproduction of pro-inflammatory cytokines, leading to acute lung injury (ALI) and systemic inflammation ("cytokine storm"). The third, final phase, the prothrombotic phase, is characterized by extensive platelet aggregation, resulting in thrombosis, coagulopathy, and multi-organ failure.

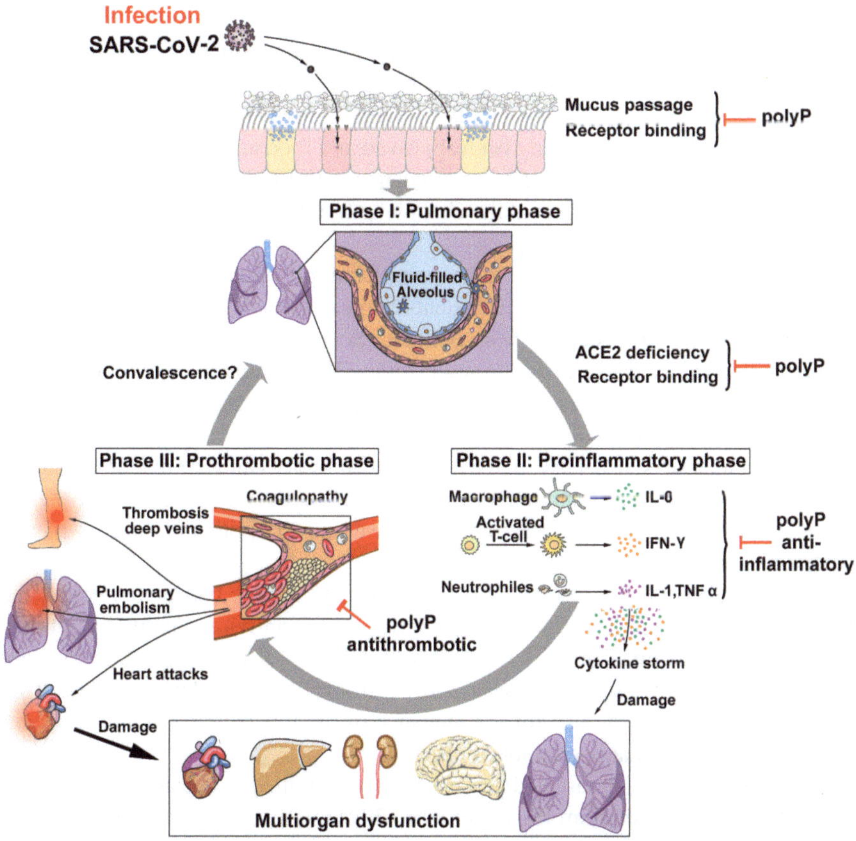

Fig. 7.8 The three phases of COVID-19 disease. The potential inhibitory effects of polyP are indicated. PolyP prevents infection of the host cells with SARS-CoV-2 and entering the pulmonary phase (phase I) by strengthening the protective function of the mucus layer (increase of mucin production) and inhibiting the attachment of the viral spike protein to the host cell ACE2 receptors in the respiratory epithelium. The pulmonary phase (phase I) is characterized by the development of pneumonia, in mild cases without hypoxia and, in more serious cases, with hypoxia requiring hospitalization. PolyP counteracts the virus-caused imbalance between the actions of the ACE and ACE2 receptors caused by the virus. In the pro-inflammatory phase (phase II), polyP is assumed to mitigate the symptoms caused by overproduction of cytokines such as interleukin-6 (IL-6), interferon-γ (IFN-γ), and tumor necrosis factor α (TNF-α) by macrophages, activated T-cells, and neutrophils, leading to cytokine storm and organ dysfunction. Finally, polyP is proposed also to act beneficially during the prothrombotic phase (phase III), due to its antithrombotic effect, through suppressing thrombosis of deep veins and embolism of lung, heart, and other organs

It is reasonable to assume that polyP can act on all three levels. During infection and in the pulmonary phase by reinforcement of the antiviral mucus shield covering the respiratory epithelium and by masking the spike RBD of the virus, thus preventing receptor binding, cell infection, and impairment of ACE2 function, in the pro-inflammatory phase by suppressing the excessive release of pro-inflammatory cytokines by interfering/inhibiting the pathways leading to their induction, and in the prothrombotic phase by substituting polyP and taking advantage of the likely antithrombotic activity of the polymer (Fig. 7.8).

PolyP is non-toxic, has already been approved as a food additive by FDA, the EU, and the JECFA (listed under E452 in accordance with EU Directive 98/72/EC) and has no proven thrombotic effect at the chain length used (40 P_i residues), in contrast to the long-chain bacterial polyP.

These properties suggest polyP to be a promising polyanion for potential clinical application in therapy of COVID-19 patients or prevention of the disease. First plans for a clinical study have recently been outlined (Schepler et al. 2021). The preferred administration appears to be via the nasal route. The respiratory epithelium of the nasal cavity and the upper respiratory tract is the primary target of infection with the virus, before it spreads to other organs.

The viral load of SARS-CoV-2 in the upper respiratory can vary in the range from 10^2 to 10^7 RNA copies/ml and higher (Zheng et al. 2020; Anand and Mayya 2020), and is a measure of the progress of the disease and the potential success of any therapeutic intervention.

Prior to infection, the virus has to overcome the mucus barrier protecting the respiratory epithelium. The functional integrity of this antiviral shield is of utmost importance for this first line of defense against the virus. It is subject to damage not only by viruses but also by bacteria and environmental pollutants, and must be permanently regenerated, also for maintenance of the mucociliary clearance mechanism, all processes that need a lot of energy.

Accordingly, the main targeted actions of polyP, applied as a drug, at the level of mucus/respiratory epithelium (Schepler et al. 2021) will be, first, strengthening the antiviral barrier function of the mucus (1) by increasing the synthesis of the main components of the mucus, the mucins such as MUC1 and MUC5AC, and (2) by exploiting the ability of polyP to form a gel-like coacervate in the presence of proteins that provides an additional layer entrapping the virus particles; second, the generation of metabolic energy (ATP), both extra- and intracellularly in the airway epithelial cells (Müller et al. 2020b), needed for these processes and to compensate a possible lack caused by cell damage or high energy consumption of virus-infected cells; and third, blocking the SARS-CoV-2 spike:host cell receptor interaction.

The development of thrombocytopenia is a characteristic symptom that is often observed in COVID-19 patients (Larsen et al. 2020). The abnormally low levels of platelets are due to an enhanced consumption of the platelets without an equivalent platelet formation (Wool and Miller 2021). It is assumed that this condition is the result of an increased activation of the platelets caused by binding of the SARS-CoV-2 spike protein (Zhang et al. 2020). The platelet activation is associated with the release of polyP stored in their dense granules, which will consequently lead, with time, to a deficiency of available polyP. The administration of polyP will help to substitute this polyP shortage. The size of the polyP used (Na-polyP$_{40}$ will be applied) is in the same size range as the soluble polyP in blood (~40 P_i units).

The polyP will be administered as an aerosol, by inhalation, e.g., in the form of a nasal spray. The most suitable formulation seems to be a combination of soluble Na-polyP and stable Ca-polyP nano/microparticles, which release the polymer after protein contact. In this formulation, the soluble form of the polyP will be immediately available, and the particulate form later, for coacervate formation to entrap the virus, whose spikes will then be masked by polyP to abolish the virus attachment to the cellular receptor.

7.13 PolyP for Prevention of SARS-CoV-2 Infection

There is an urgent need for the development of methods to prevent transmission of SARS-CoV-2. Transmission of CoV-2 and other coronaviruses occurs via inhalation of respiratory droplets or contact transmission *via* contaminated hands or surfaces and subsequent self-inoculation (Otter et al. 2016; Bahl et al. 2020; Peng et al. 2020). The respiratory droplets in the exhaled air consist both of aerosol particles (≤5 µm), which do not settle over long periods of time, and larger droplets (>5 µm) (Konda et al. 2020). They are produced by shear forces caused by the air flow and mainly contain mucin and other components of the mucus, as well as possibly virus particles. Their emission rates show a high variability; for normal breathing they amount to 0.31 particles per second and, during speaking, they are 2.77 particles per second (average numbers). During coughing, they can be very high with an average of about 10 particles per second (Asadi et al. 2020). SARS-CoV can persist on surfaces of metals, glass, or plastics for several days (Kampf et al. 2020). Therefore, the development of protective measures such as antiviral coatings and, in particular, face masks is of extreme importance, especially for health care workers who are particularly exposed to the pathogenic agent (Chua et al. 2020; Wibisono et al. 2020).

Conventional face masks according to DIN EN 149:2009 (protection classes FFP2 [retention of 94% of particles <0.5 µm] and FFP3 [99.95% of particles <0.5 µm]) that are used to prevent the spread of the virus do not have any antiviral activity (Tsai 2020). These masks are essentially based on a filtering principle whereby particles above a certain size are retained (Konda et al. 2020). The virus particles (present in aerosol particles or droplets) that remain on the mask are still infectious and potentially hazardous. In addition, the material of these masks, mostly polypropylene (Konda et al. 2020), is not biodegradable and can contaminate the environment.

The following properties are desirable for masks: Firstly, the ability to entrap and firmly bind the infectious viruses, which—mostly in aerosol droplets—reach the mask both from inside and from outside, thereby preventing virus spread and providing self-protection. Secondly, an antiviral effect either by destroying or inactivating the virus, e.g., by blocking the spike protein, thus preventing its binding to the host cell receptor if the virus is released from the mask. Thirdly, the masks should be biodegradable and not cause environmental pollution.

We succeeded to design such a mask (Müller et al. 2021a). The filter material consists of an electrospun fiber network of biodegradable polycaprolactone (PCL), into which polyP, both as a sodium salt (Na-polyP; water soluble) and in the form of polyP nanoparticles, has been incorporated. In addition, as a further component, tretinoin (retinoic acid), an inhibitor of the SARS-CoV-2 envelope (E) protein (Dey et al. 2020), a viroporin that acts as an ion channel, has been incorporated into the nanoparticles.

This polyP material has a threefold antiviral effect: (1) by entrapping the virus in a gel-like coacervate that is formed upon contact of the mask fibers with virus-containing aerosol particles, (2) by firmly binding to the SARS-CoV-2 spike protein and blocking its capability to attach to the host cell's receptor, and (3) by destroying the virus particle by a second component (retinoic acid/tretinoin) released from the nanoparticles (Müller et al. 2021a).

7.14 Future Aspects: From the Present Pandemic to Long COVID and Future Epi/pandemics

There are still a number of questions that would be of interest to be addressed in future research. The mucus covering the airway epithelial cells traps and rapidly removes foreign particles. To escape this fast clearance mechanism and reach the epithelium, the polyP if administered as a spray via the nasal route must be able to penetrate through the mucus layer. It has been proposed that the positively charged calcium ions shield the negatively charged mucin molecules, counteracting the repulsion of the polymer chains (Paz et al. 2003). This enables the condensed structure of the mucins. Scavenging of Ca^{2+} by the polyanionic polyP might disrupt the tight packaging of the mucus allowing the polymer to penetrate through the mucus layer. The effect of polyanionic polymers such as polyP on the structure and permeability of the mucus layer has to be investigated in more detail.

It should be noted that polyP if applied as nano/microparticles can be combined with other drugs incorporated into these particles. For example, amorphous nanoparticles of the magnesium salt of polyP (Mg-polyP nanoparticles) have been prepared that were loaded with quercetin or dexamethasone. It was found that the loaded nanoparticles containing the embedded drugs strongly enhanced the expression of *MUC5AC* gene in alveolar A549 cells up to levels that far exceeded the level of stimulation of each of the drugs alone (Neufurth et al. 2021). Another example are tretinoin/retinoic acid containing polyP particles, which were able to not only block host cell receptor binding of the virus, but also destroy the virus by interfering with the function of the viral envelope viroporin ion channel protein (Müller et al. 2021a).

There is increasing concern about the long-term health consequences for COVID-19 patients who can show persistent symptoms over months and possibly longer after the disease, so-called long COVID or post-COVID syndrome (Wong and Weitzer 2021). A number of clinical trials on long COVID have presently been started or are going on. Chronic fatigue syndrome (myalgic encephalomyelitis) is a condition often observed in long COVID patients. One polyanionic drug included in clinical studies is Ampligen (Rintatolimod), a mismatched dsRNA, poly(I:C_{12}U) that can be administered via the intranasal route (Strayer et al. 2020). Ampligen has been shown to be a selective TLR3 agonist (Jasani et al. 2009) and an activator of the 2-5A system (Ushijima et al. 1993; Suhadolnik et al. 1994). It also acts as an inhibitor of DNA topoisomerase I both in noninfected and HIV-1-infected H9 cells (Schröder et al. 1994a, b). In this context, it might be relevant that polyP, in yeast model, has been found to be involved in polyphosphorylation of DNA topoisomerase I, leading to an inhibition of the enzyme (Azevedo et al. 2015). Further, it has been shown that chronic fatigue syndrome in patients is associated with an impairment of the energy state of cells (Tomas et al. 2017). Based on these findings it might be interesting to consider, in future, the possible beneficial effects of polyP in therapy of chronic fatigue associated with post-COVID syndrome both with regard to its ability to act as a modulator of innate immune system and as a donor of metabolic energy, both intra- and extracellularly.

Finally, it should be emphasized that the development of drugs that reinforce the innate immune defense including, among others, the organization and composition of the mucus might also be beneficial in the fight against future pandemics caused by newly emerging zoonotic viruses. It is generally agreed that newly emerging infectious diseases are mostly zoonotic diseases that are transmitted from animals to humans and their frequency is increasing (Åsjö and Kruse 2006; Wang and Crameri 2014). Many pathogens, both bacteria and viruses, are respiratory. However, they are not known in advance and the development of drugs directed against specific targets is not possible. Therefore, the development of therapeutic strategies that can be applied to a broad range of pathogens is indicated. There is an urgent demand for such drugs. PolyP could be a promising candidate.

Acknowledgments W.E.G.M. acknowledges financial support from the European Research Council (ERC Advanced Investigator Grant No. 268476 and the three ERC-PoC grants No. 324564, No. 662486, and No. 767234). In addition, we acknowledge funding from the European Commission (grants No. 604036 and No. 311848) and the BiomaTiCS research initiative of the University Medical Center, Mainz.

References

Abrams DI, Kuno S, Wong R, Jeffords K, Nash M, Molaghan JB, Gorter R, Ueno R (1989) Oral dextran sulfate (UA001) in the treatment of the acquired immunodeficiency syndrome (AIDS) and AIDS-related complex. Ann Intern Med 110:183–188

Adusumilli NC, Zhang D, Friedman JM, Friedman AJ (2020) Harnessing nitric oxide for preventing, limiting and treating the severe pulmonary consequences of COVID-19. Nitric Oxide 103:4–8

Ahmed A, Siman-Tov G, Hall G, Bhalla N, Narayanan A (2019) Human antimicrobial peptides as therapeutics for viral infections. Viruses 11:704

Akerström S, Mousavi-Jazi M, Klingström J, Leijon M, Lundkvist A, Mirazimi A (2005) Nitric oxide inhibits the replication cycle of severe acute respiratory syndrome coronavirus. J Virol 79: 1966–1969

Akerström S, Gunalan V, Keng CT, Tan YJ, Mirazimi A (2009) Dual effect of nitric oxide on SARS-CoV replication: viral RNA production and palmitoylation of the S protein are affected. Virology 395:1–9

Allolio C, Magarkar A, Jurkiewicz P, Baxová K, Javanainen M, Mason PE, Šachl R, Cebecauer M, Hof M, Horinek D, Heinz V, Rachel R, Ziegler CM, Schröfel A, Jungwirth P (2018) Arginine-rich cell-penetrating peptides induce membrane multilamellarity and subsequently enter via formation of a fusion pore. Proc Natl Acad Sci USA 115:11923–11928

Ambrosio AL, Di Pietro SM (2017) Storage pool diseases illuminate platelet dense granule biogenesis. Platelets 28:138–146

Anand S, Mayya YS (2020) Size distribution of virus laden droplets from expiratory ejecta of infected subjects. Sci Rep 10:21174

Arepally GM (2017) Heparin-induced thrombocytopenia. Blood 129:2864–2872

Asadi S, Cappa CD, Barreda S, Wexler AS, Bouvier NM, Ristenpart WD (2020) Efficacy of masks and face coverings in controlling outward aerosol particle emission from expiratory activities. Sci Rep 10:15665

Åsjö B, Kruse H (2006) Zoonoses in the emergence of human viral diseases. Perspect Med Virol 16: 15–41

Azevedo C, Livermore T, Saiardi A (2015) Protein polyphosphorylation of lysine residues by inorganic polyphosphate. Mol Cell. 58:71–82

Aztatzi-Aguilar OG, Uribe-Ramírez M, Arias-Montaño JA, Barbier O, De Vizcaya-Ruiz A (2015) Acute and subchronic exposure to air particulate matter induces expression of angiotensin and bradykinin-related genes in the lungs and heart: Angiotensin-II type-I receptor as a molecular target of particulate matter exposure. Part Fibre Toxicol 12:17

Bae CH, Na HG, Choi YS, Song SY, Kim YD (2018) Clusterin induces MUC5AC expression *via* activation of NF-κB in human airway epithelial cells. Clin Exp Otorhinolaryngol 11:124–132

Bahl P, Doolan C, de Silva C, Chughtai AA, Bourouiba L, MacIntyre CR (2020) Airborne or droplet precautions for health workers treating COVID-19? J Infect Dis 16:jiaa189

Baker CJ, Smith SA, Morrissey JH (2018) Polyphosphate in thrombosis, hemostasis, and inflammation. Res Pract Thromb Haemost 3:18–25

Bar-On YM, Flamholz A, Phillips R, Milo R (2020) SARS-CoV-2 (COVID-19) by the numbers. Elife 9:e57309

Bianculli RH, Mase JD, Schulz MD (2020) Antiviral polymers: past approaches and future possibilities. Macromolecules 53:9158–9186

Chan JF, Kok KH, Zhu Z, Chu H, To KK, Yuan S, Yuen KY (2020) Genomic characterization of the 2019 novel human-pathogenic coronavirus isolated from a patient with atypical pneumonia after visiting Wuhan. Emerg Microbes Infect 9:221–236

Choi SH, Smith SA, Morrissey JH (2011) Polyphosphate is a cofactor for the activation of factor XI by thrombin. Blood 118:6963–6970

Chrysanthopoulou A, Kambas K, Stakos D, Mitroulis I, Mitsios A, Vidali V, Angelidou I, Bochenek M, Arelaki S, Arampatzioglou A, Galani IE, Skendros P, Couladouros EA, Konstantinides S, Andreakos E, Schäfer K, Ritis K (2017) Interferon lambda1/IL-29 and inorganic polyphosphate are novel regulators of neutrophil-driven thromboinflammation. J Pathol 243:111–122

Chua MH, Cheng W, Goh SS, Kong J, Li B, Lim JYC, Mao L, Wang S, Xue K, Yang L, Ye E, Zhang K, Cheong WCD, Tan BH, Li Z, Tan BH, Loh XJ (2020) Face masks in the new COVID-19 normal: Materials, testing, and perspectives. Research (Wash D C) 2020:7286735

Church FC, Pratt CW, Treanor RE, Whinna HC (1988) Antithrombin action of phosvitin and other phosphate-containing polyanions is mediated by heparin cofactor II. FEBS Lett 237:26–30

Cines DB, Yarovoi SV, Zaitsev SV, Lebedeva T, Rauova L, Poncz M, Arepally GM, Khandelwal S, Stepanova V, Rux AH, Cuker A, Guo C, Ocariza LM, Travers RJ, Smith SA,

Kim H, Morrissey JH, Conway EM (2016) Polyphosphate/platelet factor 4 complexes can mediate heparin-independent platelet activation in heparin-induced thrombocytopenia. Blood Adv 1:62–74
Cini N, Ball V (2014) Polyphosphates as inorganic polyelectrolytes interacting with oppositely charged ions, polymers and deposited on surfaces: fundamentals and applications. Adv Colloid Interface Sci 209:84–97
Clausen TM, Sandoval DR, Spliid CB, Pihl J, Perrett HR, Painter CD, Narayanan A, Majowicz SA, Kwong EM, McVicar RN, Thacker BE, Glass CA, Yang Z, Torres JL, Golden GJ, Bartels PL, Porell RN, Garretson AF, Laubach L, Feldman J, Yin X, Pu Y, Hauser BM, Caradonna TM, Kellman BP, Martino C, Gordts PLSM, Chanda SK, Schmidt AG, Godula K, Leibel SL, Jose J, Corbett KD, Ward AB, Carlin AF, Esko JD (2020) SARS-CoV-2 infection depends on cellular heparan sulfate and ACE2. Cell 183:1043–1057
Coutard B, Valle C, de Lamballerie X, Canard B, Seidah NG, Decroly E (2020) The spike glycoprotein of the new coronavirus 2019-nCoV contains a furin-like cleavage site absent in CoV of the same clade. Antiviral Res 176:104742
De Jager HJ, Heyns AM (1998a) Kinetics of acid-catalyzed hydrolysis of a polyphosphate in water. J Phys Chem A 102:2838–2841
De Jager H, Heyns AM (1998b) Study of the hydrolysis of sodium polyphosphate in water using Raman spectroscopy. Appl Spectrosc 52:808–814
Dey D, Borkotoky S, Banerjee M (2020) *In silico* identification of Tretinoin as a SARS-CoV-2 envelope (E) protein ion channel inhibitor. Comput Biol Med 127:104063
D'hondt C, Ponsaerts R, De Smedt H, Vinken M, De Vuyst E, De Bock M, Wang N, Rogiers V, Leybaert L, Himpens B, Bultynck G (2011) Pannexin channels in ATP release and beyond: an unexpected rendezvous at the endoplasmic reticulum. Cell Signal 23:305–316
Dinarvand P, Hassanian SM, Qureshi SH, Manithody C, Eissenberg JC, Yang L, Rezaie AR (2014) Polyphosphate amplifies proinflammatory responses of nuclear proteins through interaction with receptor for advanced glycation end products and $P2Y_1$ purinergic receptor. Blood 123: 935–945
Docampo R, de Souza W, Miranda K, Rohloff P, Moreno SN (2005) Acidocalcisomes - conserved from bacteria to man. Nat Rev Microbiol 3:251–261
Domingo LR, Ríos-Gutierrez M, Pérez P (2017) A molecular electron density theory study of the [3 + 2] cycloaddition reaction of nitrones with strained allenes. RSC Adv 7:26879–26887
Dong N, Yang X, Ye L, Chen K, Chan EWC, Yang M, Chen S (2020) Genomic and protein structure modelling analysis depicts the origin and infectivity of 2019-nCoV, a new coronavirus which caused a pneumonia outbreak in Wuhan, China, bioRxiv 2020. https://doi.org/10.1101/2020.01.20.913368
Donoghue M, Hsieh F, Baronas E, Godbout K, Gosselin M, Stagliano N, Donovan M, Woolf B, Robison K, Jeyaseelan R, Breitbart RE, Acton S (2000) A novel angiotensin-converting enzyme-related carboxypeptidase (ACE2) converts angiotensin I to angiotensin 1-9. Circ Res 87:E1–E9
Droguett K, Rios M, Carreño DV, Navarrete C, Fuentes C, Villalón M, Barrera NP (2017) An autocrine ATP release mechanism regulates basal ciliary activity in airway epithelium. J Physiol 595:4755–4767
Eckert DM, Kim PS (2001) Mechanisms of viral membrane fusion and its inhibition. Annu Rev Biochem 70:777–810
Ferrucci V, Kong DY, Asadzadeh F, Marrone L, Boccia A, Siciliano R, Criscuolo G, Anastasio C, Quarantelli F, Comegna M, Pisano I, Passariello M, Iacobucci I, Monica RD, Izzo B, Cerino P, Fusco G, Viscardi M, Brandi S, Pierri BM, Borriello G, Tiberio C, Atripaldi L, Bianchi M, Paolella G, Capoluongo E, Castaldo G, Chiariotti L, Monti M, De Lorenzo C, Yun KS, Pascarella S, Cheong JH, Kim HY, Zollo M (2021) Long-chain polyphosphates impair SARS-CoV-2 infection and replication. Sci Signal 14:eabe5040
Fitzgerald KA, Kagan JC (2020) Toll-like receptors and the control of immunity. Cell 180:1044–1066

Ganz T (2002) Antimicrobial polypeptides in host defense of the respiratory tract. J Clin Invest 109: 693–697

Garcia-Del-Barco D, Risco-Acevedo D, Berlanga-Acosta J, Martos-Benítez FD, Guillén-Nieto G (2021) Revisiting pleiotropic effects of type I interferons: rationale for its prophylactic and therapeutic use against SARS-CoV-2. Front Immunol 12:655528

Gasparini G, Bang EK, Montenegro J, Matile S (2015) Cellular uptake: lessons from supramolecular organic chemistry. Chem Commun (Camb) 51:10389–10402

Gerasimaitė R, Sharma S, Desfougères Y, Schmidt A, Mayer A (2014) Coupled synthesis and translocation restrains polyphosphate to acidocalcisome-like vacuoles and prevents its toxicity. J Cell Sci 127:5093–5104

Gil C, Ginex T, Maestro I, Nozal V, Barrado-Gil L, Cuesta-Geijo MÁ, Urquiza J, Ramírez D, Alonso C, Campillo NE, Martinez A (2020) COVID-19: Drug targets and potential treatments. J Med Chem 63:12359–12386

Ginsburg I, Fibach E (2021) Polycations and polyanions in SARS-CoV-2 infection. Med Hypotheses 146:110470

Gordon LM, Waring AJ, Curtain CC, Kirkpatrick A, Leung C, Faull K, Mobley PW (1995) Antivirals that target the amino-terminal domain of HIV type 1 glycoprotein 41. AIDS Res Hum Retroviruses 11:677–686

Gozzo L, Viale P, Longo L, Vitale DC, Drago F (2020) The potential role of heparin in patients with COVID-19: beyond the anticoagulant effect. A review. Front Pharmacol 11:1307

Guzman-Aranguez A, Argüeso P (2010) Structure and biological roles of mucin-type *O*-glycans at the ocular surface. Ocul Surf 8:8–17

Hansson GC (2019) Mucus and mucins in diseases of the intestinal and respiratory tracts. J Intern Med 285:479–490

Harada K, Shiba T, Doi K, Morita K, Kubo T, Makihara Y, Piattelli A, Akagawa Y (2013) Inorganic polyphosphate suppresses lipopolysaccharide-induced inducible nitric oxide synthase (iNOS) expression in macrophages. PLoS One 8:e74650

Harrison SC (2015) Viral membrane fusion. Virology 479-480:498–507

Hiti AL, Davis AR, Nayak DP (1981) Complete sequence analysis shows that the hemagglutinins of the H0 and H2 subtypes of human influenza virus are closely related. Virology 111:113–124

Hoffmann M, Hofmann-Winkler H, Pöhlmann S (2018) Priming time: how cellular proteases arm coronavirus spike proteins. Activat Virus Host Proteases 16:71–98

Hoffmann M, Kleine-Weber H, Krüger N, Müller M, Drosten C, Pöhlmann S (2020a) The novel coronavirus 2019 (2019-nCoV) uses the SARS coronavirus receptor ACE2 and the cellular protease TMPRSS2 for entry into target cells. bioRxiv [Preprint]. https://doi.org/10.1101/2020.01.31.929042

Hoffmann M, Kleine-Weber H, Schroeder S, Krüger N, Herrler T, Erichsen S, Schiergens TS, Herrler G, Wu NH, Nitsche A, Müller MA, Drosten C, Pöhlmann S (2020b) SARS-CoV-2 cell entry depends on ACE2 and TMPRSS2 and is blocked by a clinically proven protease inhibitor. Cell 181:271–280

Hoffmann M, Kleine-Weber H, Pöhlmann S (2020c) A Multibasic cleavage site in the spike protein of SARS-CoV-2 is essential for infection of human lung cells. Mol Cell 78:779–784

Holmström KM, Marina N, Baev AY, Wood NW, Gourine AV, Abramov AY (2013) Signalling properties of inorganic polyphosphate in the mammalian brain. Nat Commun 4:1362

Hosoya M, Balzarini J, Shigeta S, De Clercq E (1991) Differential inhibitory effects of sulfated polysaccharides and polymers on the replication of various myxoviruses and retroviruses, depending on the composition of the target amino acid sequences of the viral envelope glycoproteins. Antimicrob Agents Chemother 35:2515–2520

Huang Y, Yang C, Xu XF, Xu W, Liu SW (2020) Structural and functional properties of SARS-CoV-2 spike protein: potential antivirus drug development for COVID-19. Acta Pharmacol Sin 41:1141–1149

Hulswit RJ, de Haan CA, Bosch BJ (2016) Coronavirus spike protein and tropism changes. Adv Virus Res 96:29–57

Hussman JP (2020) Cellular and molecular pathways of COVID-19 and potential points of therapeutic intervention. Front Pharmacol 11:1169

Ison MG, Hayden FG (2017) Antiviral agents against respiratory viruses. Infect Dis 2:1318–1326

Itakura E, Chiba M, Murata T, Matsuura A (2020) Heparan sulfate is a clearance receptor for aberrant extracellular proteins. J Cell Biol 219:e201911126

Jasani B, Navabi H, Adams M (2009) Ampligen: a potential toll-like 3 receptor adjuvant for immunotherapy of cancer. Vaccine 27:3401–3404

Kaler L, Iverson E, Bader S, Song D, Scull MA, Duncan GA (2020) Influenza A virus diffusion through mucus gel networks. bioRxiv preprint. https://doi.org/10.1101/2020.08.14.251132

Kampf G, Todt D, Pfaender S, Steinmann E (2020) Persistence of coronaviruses on inanimate surfaces and their inactivation with biocidal agents. J Hosp Infect 104:246–251

Katchman BJ, Smith HE (1958) Diffusion of synthetic and natural polyphosphates. Arch Biochem Biophys 76:396–402

Kawai T, Akira S (2007) Signaling to NF-kappaB by Toll-like receptors. Trends Mol Med 13:460–469

Kawase M, Kataoka M, Shirato K, Matsuyama S (2019) Biochemical analysis of coronavirus spike glycoprotein conformational intermediates during membrane fusion. J Virol 93:e00785–e00719

Kesimer M, Makhov AM, Griffith JD, Verdugo P, Sheehan JK (2010) Unpacking a gel-forming mucin: a view of MUC5B organization after granular release. Am J Physiol Lung Cell Mol Physiol 298:L15–L22

Kifle ZD, Ayele AG, Enyew EF (2021) Drug repurposing approach, potential drugs, and novel drug targets for COVID-19 treatment. J Environ Public Health 2021:6631721

Kim SY, Jin W, Sood A, Montgomery DW, Grant OC, Fuster MM, Fu L, Dordick JS, Woods RJ, Zhang F, Linhardt RJ (2020) Characterization of heparin and severe acute respiratory syndrome-related coronavirus 2 (SARS-CoV-2) spike glycoprotein binding interactions. Antiviral Res 181:104873

Kirk-Othmer Encyclopedia of Chemical Technology (1991) Phosphoric acid and phosphates, 4th edn, vol 18, pp 669–718

Konda A, Prakash A, Moss GA, Schmoldt M, Grant GD, Guha S (2020) Aerosol filtration efficiency of common fabrics used in respiratory cloth masks. ACS Nano 14:6339–6347

Kondo Y, Kuroda Y, Nakagawa T, Shimizu N, Murata M (2019) Assessment of *in vitro* effects of direct thrombin inhibitors and activated factor X inhibitors through clot waveform analysis. J Clin Pathol 72:244–250

Kornberg A, Rao NN, Ault-Riché D (1999) Inorganic polyphosphate: a molecule of many functions. Annu Rev Biochem 68:89–125

Kristiansen H, Scherer CA, McVean M, Iadonato SP, Vends S, Thavachelvam K, Steffensen TB, Horan KA, Kuri T, Weber F, Paludan SR, Hartmann R (2010) Extracellular 2′-5′ oligoadenylate synthetase stimulates RNase L-independent antiviral activity: a novel mechanism of virus-induced innate immunity. J Virol 84:11898–11904

Kulaev IS, Vagabov V, Kulakovskaya T (2004) The biochemistry of inorganic polyphosphates. Wiley, Chichester

Kuusksalu A, Truve E, Aaspollu A, Kelve M, Scheffer U, Müller WEG, Schröder HC (1995) Impairment of intracellular antiviral defense with age: age-dependent changes in expression of interferon-induced and double-stranded RNA-activated 2-5A synthetase in rat. Mech Ageing Dev 78:103–115

Kyte J, Doolittle RF (1982) A simple method for displaying the hydropathic character of a protein. J Mol Biol 157:105–132

Lai SK, Wang YY, Hanes J (2009) Mucus-penetrating nanoparticles for drug and gene delivery to mucosal tissues. Adv Drug Deliv Rev 61:158–171

Lan J, Ge J, Yu J, Shan S, Zhou H, Fan S, Zhang Q, Shi X, Wang Q, Zhang L, Wang X (2020) Structure of the SARS-CoV-2 spike receptor-binding domain bound to the ACE2 receptor. Nature 581:215–220

Larsen JB, Pasalic L, Hvas AM (2020) Platelets in coronavirus disease 2019. Semin Thromb Hemost 46:823–825

Lee C, Choi WJ (2021) Overview of COVID-19 inflammatory pathogenesis from the therapeutic perspective. Arch Pharm Res 44:99–116

Lee RM, Hartman PA, Stahr HM, Olson DG, Williams FD (1994) Antibacterial mechanism of long-chain polyphosphates in *Staphylococcus aureus*. J Food Prot 57:289–294

Lempart J, Tse E, Lauer JA, Ivanova MI, Sutter A, Yoo N, Huettemann P, Southworth D, Jakob U (2019) Mechanistic insights into the protective roles of polyphosphate against amyloid cytotoxicity. Life Sci Alliance 2:e201900486

Li Y, Renner DM, Comar CE, Whelan JN, Reyes HM, Cardenas-Diaz FL, Truitt R, Tan LH, Dong B, Alysandratos KD, Huang J, Palmer JN, Adappa ND, Kohanski MA, Kotton DN, Silverman RH, Yang W, Morrisey EE, Cohen NA, Weiss SR (2021) SARS-CoV-2 induces double-stranded RNA-mediated innate immune responses in respiratory epithelial-derived cells and cardiomyocytes. Proc Natl Acad Sci USA 118:e2022643118

Lippmann F (1951) In phosphorus metabolism. John Hopkins Press, Baltimore, pp 521–525

Liu T, Luo S, Libby P, Shi GP (2020) Cathepsin L-selective inhibitors: a potentially promising treatment for COVID-19 patients. Pharmacol Ther 213:107587

Lorencová E, Vltavská P, Budinský P, Koutný M (2012) Antibacterial effect of phosphates and polyphosphates with different chain length. J Environ Sci Health A Tox Hazard Subst Environ Eng 47:2241–2245

Lorentsen KJ, Hendrix CW, Collins JM, Kornhauser DM, Petty BG, Klecker RW, Flexner C, Eckel RH, Lietman PS (1989) Dextran sulfate is poorly absorbed after oral administration. Ann Intern Med 111:561–566

Lorenz B, Leuck J, Köhl D, Müller WEG, Schröder HC (1997) Anti-HIV-1 activity of inorganic polyphosphates. J Acquir Immune Defic Syndr Hum Retrovirol 14:110–118

Lu R, Zhao X, Li J, Niu P, Yang B, Wu H, Wang W, Song H, Huang B, Zhu N, Bi Y, Ma X, Zhan F, Wang L, Hu T, Zhou H, Hu Z, Zhou W, Zhao L, Chen J, Meng Y, Wang J, Lin Y, Yuan J, Xie Z, Ma J, Liu WJ, Wang D, Xu W, Holmes EC, Gao GF, Wu G, Chen W, Shi W, Tan W (2020) Genomic characterization and epidemiology of 2019 novel coronavirus: implications for virus origins and receptor binding. Lancet 395:565–574

Lüscher-Mattli M (2000) Polyanions – a lost chance in the fight against HIV and other virus diseases? Antivir Chem Chemother 11:249–259

Lüscher-Mattli M, Glück R (1990) Dextran sulfate inhibits the fusion of influenza virus with model membranes, and suppresses influenza virus replication *in vivo*. Antiviral Res 14:39–50

Lüscher-Mattli M, Glück R, Kempf C, Zanoni-Grassi M (1993) A comparative study of the effect of dextran sulfate on the fusion and the *in vitro* replication of influenza A and B, Semliki Forest, vesicular stomatitis, rabies, Sendai, and mumps virus. Arch Virol 130:317–326

Magro G (2020) SARS-CoV-2 and COVID-19: Is interleukin-6 (IL-6) the 'culprit lesion' of ARDS onset? What is there besides Tocilizumab? SGP130Fc. Cytokine 2:100029

Majling J, Hanic F (1980) Phase chemistry of condensed phosphates. Top Phosphorus Chem 10: 341–502

Matsuyama T, Kubli SP, Yoshinaga SK, Pfeffer K, Mak TW (2020) An aberrant STAT pathway is central to COVID-19. Cell Death Differ 27:3209–3225

McCarthy WJ, Adamowicz L, Saint-Martin H, Ortega-Blake I (2002) A comparative ab initio study of the isomerizations and hydrolyses of neutral and anionic M-pyrophosphate complexes, with M = Ca, Zn. Rev Soc Quím Mex 46:145–158

Melkamu T, Kita H, O'Grady SM (2013) TLR3 activation evokes IL-6 secretion, autocrine regulation of Stat3 signaling and TLR2 expression in human bronchial epithelial cells. J Cell Commun Signal 7:109–118

Meyer D, Sielaff F, Hammami M, Böttcher-Friebertshäuser E, Garten W, Steinmetzer T (2013) Identification of the first synthetic inhibitors of the type II transmembrane serine protease TMPRSS2 suitable for inhibition of influenza virus activation. Biochem J 452:331–343

Millet JK, Whittaker GR (2018) Physiological and molecular triggers for SARS-CoV membrane fusion and entry into host cells. Virology 517:3–8

Mishra A, Lai GH, Schmidt NW, Sun VZ, Rodriguez AR, Tong R, Tang L, Cheng J, Deming TJ, Kamei DT, Wong GC (2011) Translocation of HIV TAT peptide and analogues induced by multiplexed membrane and cytoskeletal interactions. Proc Natl Acad Sci USA 108:16883–16888

Mitchell JB, Thornton JM, Singh J, Price SL (1992) Towards an understanding of the arginine-aspartate interaction. J Mol Biol 226:251–262

Mitsuya H, Looney DJ, Kuno S, Ueno R, Wong-Staal F, Broder S (1988) Dextran sulfate suppression of viruses in the HIV family: inhibition of virion binding to CD4+ cells. Science 240:646–649

Mohan P, Schols D, Baba M, De Clercq E (1992) Sulfonic acid polymers as a new class of human immunodeficiency virus inhibitors. Antiviral Res 18:139–150

Moon JH, Park JH, Lee JY (2011) Antibacterial action of polyphosphate on *Porphyromonas gingivalis*. Antimicrob Agents Chemother 55:806–812

Morrissey JH, Choi SH, Smith SA (2012) Polyphosphate: an ancient molecule that links platelets, coagulation, and inflammation. Blood 119:5972–5979

Müller WEG, Schröder HC, Zahn RK, Dose K (1980) Degradation of 2′-5′-linked oligoriboadenylates by 3′-exoribonuclease and 5′-nucleotidase from calf thymus. Hoppe Seylers Z Physiol Chem 361:469–472

Müller WEG, Okamoto T, Reuter P, Ugarkovic D, Schröder HC (1990) Functional characterization of Tat protein from human immunodeficiency virus. Evidence that Tat links viral RNAs to nuclear matrix. J Biol Chem 265:3803–3808

Müller F, Mutch NJ, Schenk WA, Smith SA, Esterl L, Spronk HM, Schmidbauer S, Gahl WA, Morrissey JH, Renné T (2009) Platelet polyphosphates are proinflammatory and procoagulant mediators *in vivo*. Cell 139:1143–1156

Müller WEG, Wang XH, Schröder HC (eds) (2013) Biomedical inorganic polymers: bioactivity and applications of natural and synthetic polymeric inorganic molecules. Springer, Berlin, pp 1–303

Müller WEG, Neufurth M, Ackermann M, Tolba E, Korzhev M, Wang S, Feng Q, Schröder HC, Wang XL (2017a) Bifunctional dentifrice: amorphous polyphosphate a regeneratively active sealant with potent anti-*Streptococcus mutans* activity. Dent Mater 33:753–764

Müller WEG, Wang S, Ackermann M, Neufurth M, Steffen R, Mecja E, Muñoz-Espí R, Feng Q, Schröder HC, Wang XH (2017b) Rebalancing β-amyloid-Induced decrease of ATP level by amorphous nano/micro polyphosphate: Suppression of the neurotoxic effect of amyloid β-protein fragment 25-35. Int J Mol Sci 18:2154

Müller WEG, Wang S, Neufurth M, Kokkinopoulou M, Feng Q, Schröder HC, Wang XH (2017c) Polyphosphate as a donor of high-energy phosphate for the synthesis of ADP and ATP. J Cell Sci 130:2747–2756

Müller WEG, Neufurth M, Wang S, Ackermann M, Muñoz-Espí R, Feng Q, Lu Q, Schröder HC, Wang XH (2018a) Amorphous, smart, and bioinspired polyphosphate nano/microparticles: a biomaterial for regeneration and repair of osteo-articular impairments *in-situ*. Int J Mol Sci 19:427

Müller WEG, Wang SF, Tolba E, Neufurth M, Ackermann M, Muñoz-Espí R, Lieberwirth I, Glasser G, Schröder HC, Wang XH (2018b) Transformation of amorphous polyphosphate nanoparticles into coacervate complexes: an approach for the encapsulation of mesenchymal stem cells. Small 14:e1801170

Müller WEG, Schröder HC, Wang XH (2019) Inorganic polyphosphates as storage for and generator of metabolic energy in the extracellular matrix. Chem Rev 119:12337–12374

Müller WEG, Neufurth M, Schepler H, Wang SF, Tolba E, Schröder HC, Wang XH (2020a) The biomaterial polyphosphate blocks stoichiometric binding of the SARS-CoV-2 S-protein to the cellular ACE2 receptor. Biomater Sci 8:6603–6610

Müller WEG, Neufurth M, Wang SF, Tan RW, Schröder HC, Wang XH (2020b) Morphogenetic (mucin expression) as well as potential anti-corona viral activity of the marine secondary metabolite polyphosphate on A549 cells. Mar Drugs 18:639

Müller WEG, Neufurth M, Lieberwirth I, Muñoz-Espí R, Wang S, Schröder HC, Wang XH (2021a) Triple-target stimuli-responsive anti-COVID-19 face mask with physiological virus-inactivating agents. Biomater Sci 9:6052–6063

Müller WEG, Schröder HC, Neufurth M, Wang XH (2021b) An unexpected biomaterial against SARS-CoV-2: bio-polyphosphate blocks the docking of the viral spike protein with the cell receptor. Mater Today 51:504–524

Mulloy B, Hogwood J, Gray E, Lever R, Page CP (2016) Pharmacology of heparin and related drugs. Pharmacol Rev 68:76–141

Mycroft-West C, Su D, Elli S, Guimond S, Miller G, Turnbull J, Yates E, Guerrini M, Fernig D, Lima M, Skidmore M (2020a) The 2019 coronavirus (SARS-CoV-2) surface protein (Spike) S1 receptor binding domain undergoes conformational change upon heparin binding. bioRxiv 2020. https://doi.org/10.1101/2020.02.29.971093

Mycroft-West CJ, Su D, Pagani I, Rudd TR, Elli S, Gandhi NS, Guimond SE, Miller GJ, Meneghetti MCZ, Nader HB, Li Y, Nunes QM, Procter P, Mancini N, Clementi M, Bisio A, Forsyth NR, Ferro V, Turnbull JE, Guerrini M, Fernig DG, Vicenzi E, Yates EA, Lima MA, Skidmore MA (2020b) Heparin inhibits cellular invasion by SARS-CoV-2: Structural dependence of the interaction of the spike S1 receptor-binding domain with heparin. Thromb Haemost 120:1700–1715

Mycroft-West CJ, Su DH, Li Y, Guimond SE, Rudd TR, Elli S, Miller G, Nunes QM, Procter P, Bisio A, Forsyth NR, Turnbull JE, Guerrini M, Fernig DG, Yates EA, Lima MA, Skidmore MA (2020c) Glycosaminoglycans induce conformational change in the SARS-CoV-2 spike S1 receptor binding domain. bioRxiv. https://doi.org/10.1101/2020.04.29.068767

Nakashima H, Yoshida O, Tochikura TS, Yoshida T, Mimura T, Kido Y, Motoki Y, Kaneko Y, Uryu T, Yamamoto N (1987) Sulfation of polysaccharides generates potent and selective inhibitors of human immunodeficiency virus infection and replication *in vitro*. Jpn J Cancer Res 78:1164–1168

Neufurth M, Wang XH, Tolba E, Lieberwirth I, Wang S, Schröder HC, Müller WEG (2020) The inorganic polymer, polyphosphate, blocks binding of SARS-CoV-2 spike protein to ACE2 receptor at physiological concentrations. Biochem Pharmacol 182:114215

Neufurth M, Wang XH, Wang SF, Schröder HC, Müller WEG (2021) Caged dexamethasone/quercetin nanoparticles, formed of the morphogenetic active inorganic polyphosphate, are strong inducers of MUC5AC. Mar Drugs 19:64

Obritsch JA, Ryu D, Lampila LE, Bullerman LB (2008) Antibacterial effects of long-chain polyphosphates on selected spoilage and pathogenic bacteria. J Food Prot 71:1401–1405

Olmsted SS, Padgett JL, Yudin AI, Whaley KJ, Moench TR, Cone RA (2001) Diffusion of macromolecules and virus-like particles in human cervical mucus. Biophys J 81:1930–1937

Otter JA, Donskey C, Yezli S, Douthwaite S, Goldenberg SD, Weber DJ (2016) Transmission of SARS and MERS coronaviruses and influenza virus in healthcare settings: the possible role of dry surface contamination. J Hosp Infect 92:235e50

Ou X, Liu Y, Lei X, Li P, Mi D, Ren L, Guo L, Guo R, Chen T, Hu J, Xiang Z, Mu Z, Chen X, Chen J, Hu K, Jin Q, Wang J, Qian Z (2020) Characterization of spike glycoprotein of SARS-CoV-2 on virus entry and its immune cross-reactivity with SARS-CoV. Nat Commun 11:1620

Ozaki Y, Wakui M, Fujimori Y, Oka S, Nakamura S, Kondo Y, Nakagawa T, Katagiri H, Murata M (2021) *In vitro* unexpected effects of polyphosphates observed through activated partial thromboplastin time-based clot waveform analysis. Int J Lab Hematol 43:O234–O237

Paital B, Agrawal PK (2021) Air pollution by NO_2 and PM2.5 explains COVID-19 infection severity by overexpression of angiotensin-converting enzyme 2 in respiratory cells: a review. Environ Chem Lett 19:25–42

Pantos A, Tsogas I, Paleos CM (2008) Guanidinium group: a versatile moiety inducing transport and multicompartmentalization in complementary membranes. Biochim Biophys Acta 1778: 811–823
Patra R, Chandra Das N, Mukherjee S (2020) Targeting human TLRs to combat COVID-19: a solution? J Med Virol 10:1002
Patthy L, Smith EL (1975) Reversible modification of arginine residues. Application to sequence studies by restriction of tryptic hydrolysis to lysine residues. J Biol Chem 250:557–564
Paz HB, Tisdale AS, Danjo Y, Spurr-Michaud SJ, Argüeso P, Gipson IK (2003) The role of calcium in mucin packaging within goblet cells. Exp Eye Res. 77:69–75
Peng X, Xu X, Li Y, Cheng L, Zhou X, Ren B (2020) Transmission routes of 2019-nCoV and controls in dental practice. Int J Oral Sci 12:9
Petrou G, Crouzier T (2018) Mucins as multifunctional building blocks of biomaterials. Biomater Sci 6:2282–2297
Picher M, Burch LH, Hirsh AJ, Spychala J, Boucher RC (2003) Ecto 5′-nucleotidase and nonspecific alkaline phosphatase. Two AMP-hydrolyzing ectoenzymes with distinct roles in human airways. J Biol Chem 278:13468–13479
Pozzer A, Dominici F, Haines A, Witt C, Münzel T, Lelieveld J (2020) Regional and global contributions of air pollution to risk of death from COVID-19. Cardiovasc Res 116:2247–2253
Qian J, Mills DA, Geren L, Wang K, Hoganson CW, Schmidt B, Hiser C, Babcock GT, Durham B, Millett F, Ferguson-Miller S (2004) Role of the conserved arginine pair in proton and electron transfer in cytochrome C oxidase. Biochemistry 43:5748–5756
Raghunathan V, Zilberman-Rudenko J, Olson SR, Lupu F, McCarty OJT, Shatzel JJ (2019) The contact pathway and sepsis. Res Pract Thromb Haemost 3:331–339
Rambaut A, Loman N, Pybus O, Barclay W, Barrett J, Carabelli A, Connor T, Peacock T, Robertson DL, Volz E, on behalf of COVID-19 Genomics Consortium UK (2020) Preliminary genomic characterisation of an emergent SARS-CoV-2 lineage in the UK defined by a novel set of spike mutations. Preprint at https://virological.org/t/preliminary-genomic-characterisation-of-an-emergent-sars-cov-2-lineage-in-the-uk-defined-by-a-novel-set-of-spike-mutations/563
Rao NN, Gomez-Garcia MR, Kornberg A (2009) Inorganic polyphosphate: Essential for growth and survival. Annu Rev Biochem 78:605–647
Ren Y, Shu T, Wu D, Mu J, Wang C, Huang M, Han Y, Zhang XY, Zhou W, Qiu Y, Zhou X (2020) The ORF3a protein of SARS-CoV-2 induces apoptosis in cells. Cell Mol Immunol 17:881–883
Ridley C, Thornton DJ (2018) Mucins: the frontline defence of the lung. Biochem Soc Trans 46: 1099–1106
Roewe J, Stavrides G, Strueve M, Sharma A, Marini F, Mann A, Smith SA, Kaya Z, Strobl B, Mueller M, Reinhardt C, Morrissey JH, Bosmann M (2020) Bacterial polyphosphates interfere with the innate host defense to infection. Nat Commun 11:4035
Roger P, Gascard JP, Bara J, de Montpreville VT, Yeadon M, Brink C (2000) ATP induced MUC5AC release from human airways *in vitro*. Mediators Inflamm 9:277–284
Sa Ribero M, Jouvenet N, Dreux M, Nisole S (2020) Interplay between SARS-CoV-2 and the type I interferon response. PLoS Pathog 16:e1008737
Schepler H, Wang X, Neufurth M, Wang S, Schröder HC, Müller WEG (2021) The therapeutic potential of inorganic polyphosphate: a versatile physiological polymer to control coronavirus disease (COVID-19). Theranostics 11:6193–6213
Schmidt N, Mishra A, Lai GH, Wong GC (2010) Arginine-rich cell-penetrating peptides. FEBS Lett 584:1806–1813
Schols D, De Clercq E, Balzarini J, Baba M, Witvrouw M, Hosoya M, Andrei G, Snoek R, Neyts J, Pauwels R, Nagy M, Györgyi-Edelényi J, Machowich R, Horváth I, Low M, Görög S (1990) Sulphated polymers are potent and selective inhibitors of various enveloped viruses, including herpes simplex virus, cytomegalovirus, vesicular stomatitis virus, respiratory syncytial virus, and toga-, arena- and retroviruses. Antiviral Chem Chemother 1:233–240
Schröder HC, Zahn RK, Dose K, Müller WEG (1980) Purification and characterization of a poly (A)-specific exoribonuclease from calf thymus. J Biol Chem 255:4535–4538

Schröder HC, Wenger R, Kuchino Y, Müller WEG (1989) Modulation of nuclear matrix-associated (2′-5′)oligoadenylate metabolism and ribonuclease L activity in H9 cells by human immunodeficiency virus. J Biol Chem 264:5669–5673

Schröder HC, Ugarkovic D, Merz H, Kuchino Y, Okamoto T, Müller WEG (1990a) Protection of HeLa-T4+ cells against human immunodeficiency virus (HIV) infection after stable transfection with HIV LTR-2′,5′-oligoadenylate synthetase hybrid gene. FASEB J 4:3124–3130

Schröder HC, Ugarkovic D, Wenger R, Reuter P, Okamoto T, Müller WEG (1990b) Binding of tat protein to TAR region of human immunodeficiency virus type 1 blocks TAR-mediated activation of (2′-5′)oligoadenylate synthetase. AIDS Res Hum Retroviruses 6:659–672

Schröder HC, Suhadolnik RJ, Pfleiderer W, Charubala R, Müller WEG (1992) (2′-5′)oligoadenylate and intracellular immunity against retrovirus infection. Int J Biochem 24:55–63

Schröder HC, Kelve M, Schäcke H, Pfleiderer W, Charubala R, Suhadolnik RJ, Müller WEG (1994a) Inhibition of DNA topoisomerase I activity by 2′,5′-oligoadenylates and mismatched double-stranded RNA in uninfected and HIV-1-infected H9 cells. Chem Biol Interact 90:169–183

Schröder HC, Kelve M, Müller WEG (1994b) The 2-5A system and HIV infection. In: Biological Response Modifiers - Interferons, Double-Stranded RNA and 2′,5′-Oligoadenylates (Müller WEG, Schröder HC, eds). Prog Mol Subcell Biol 14:176–197

Schröder HC, Lorenz B, Kurz L, Müller WEG (1999) Inorganic polyP in eukaryotes: enzymes, metabolism and function. In: Inorganic polyphosphates – biochemistry, biology, biotechnology (Schröder HC, Müller WEG, eds). Prog Mol Subcell Biol 23:45–81

Schröder HC, Wang XH, Müller WEG (2019) Amorphous polyphosphate nanoparticles: application of the morphogenetically active inorganic polymer for personalized tissue regeneration. J Phys D Appl Phys 52:363001

Schröder HC, Wang XH, Müller WEG (2020) Biomimetic routes to micro/nanofabrication: morphogenetically active high-energy inorganic polyphosphate nano/microparticles. In: Guarino V, Focarete ML, Pisignano D (eds) Advances in nanostructured materials and nanopatterning technologies. Elsevier, São Paulo, pp 83–113

Semlyen JA, Flory PJ (1966) Analysis of the unperturbed dimensions of the linear polyphosphate chain. Trans Faraday Soc 1966:2622–2631

Shang J, Wan Y, Luo C, Ye G, Geng Q, Auerbach A, Li F (2020) Cell entry mechanisms of SARS-CoV-2. Proc Natl Acad Sci USA 117:11727–11734

Shin CY, Kim KC, Lee WJ, Jo MJ, Park KH, Dalby R, Ko KH (2000) Inhaled ATP causes mucin release from goblet cells of intact rats. Exp Lung Res 26:1–11

Shishikura Y, Koarai A, Aizawa H, Yamaya M, Sugiura H, Watanabe M, Hashimoto Y, Numakura T, Makiguti T, Abe K, Yamada M, Kikuchi T, Hoshikawa Y, Okada Y, Ichinose M (2016) Extracellular ATP is involved in dsRNA-induced MUC5AC production via P2Y2R in human airway epithelium. Respir Res 17:121

Simmons G, Zmora P, Gierer S, Heurich A, Pöhlmann S (2013) Proteolytic activation of the SARS-coronavirus spike protein: cutting enzymes at the cutting edge of antiviral research. Antiviral Res 100:605–614

Smith SA, Choi SH, Davis-Harrison R, Huyck J, Boettcher J, Rienstra CM, Morrissey JH (2010) Polyphosphate exerts differential effects on blood clotting, depending on polymer size. Blood 116:4353–4359

Sol IS, Kim YH, Park YA, Lee KE, Hong JY, Kim MN, Kim YS, Oh MS, Yoon SH, Kim MJ, Kim KW, Sohn MH, Kim KE (2016) Relationship between sputum clusterin levels and childhood asthma. Clin Exp Allergy 46:688–695

Strayer DR, Young D, Mitchell WM (2020) Effect of disease duration in a randomized Phase III trial of rintatolimod, an immune modulator for myalgic encephalomyelitis/chronic fatigue syndrome. PLoS One 15:e0240403

Suckau D, Mak M, Przybylski M (1992) Protein surface topology-probing by selective chemical modification and mass spectrometric peptide mapping. Proc Natl Acad Sci USA 89:5630–5634

Suess PM, Chinea LE, Pilling D, Gomer RH (2019) Extracellular polyphosphate promotes macrophage and fibrocyte differentiation, inhibits leukocyte proliferation, and acts as a chemotactic agent for neutrophils. J Immunol 203:493–499

Suhadolnik RJ, Reichenbach NL, Hitzges P, Sobol RW, Peterson DL, Henry B, Ablashi DV, Müller WEG, Schröder HC, Carter WA, Strayer DR (1994) Upregulation of the 2-5A synthetase/RNase L antiviral pathway associated with chronic fatigue syndrome. Clin Infect Dis 18 (Suppl 1):S96–104

Tai W, He L, Zhang X, Pu J, Voronin D, Jiang S, Zhou Y, Du L (2020) Characterization of the receptor-binding domain (RBD) of 2019 novel coronavirus: implication for development of RBD protein as a viral attachment inhibitor and vaccine. Cell Mol Immunol 17:613–620

Tan GT, Wickramasinghe A, Verma S, Hughes SH, Pezzuto JM, Baba M, Mohan P (1993) Sulfonic acid polymers are potent inhibitors of HIV-1 induced cytopathogenicity and the reverse transcriptases of both HIV-1 and HIV-2. Biochim Biophys Acta 1181:183–188

Tang T, Bidon M, Jaimes JA, Whittaker GR, Daniel S (2020) Coronavirus membrane fusion mechanism offers a potential target for antiviral development. Antiviral Res 178:104792

Tay MZ, Poh CM, Rénia L, MacAry PA, Ng LFP (2020) The trinity of COVID-19: immunity, inflammation and intervention. Nat Rev Immunol 20:363–374

Thilo E, Wicker W (1957) Zur Chemie der kondersierten Phosphate und Arsenate. XVIII. Anionenhydrolyse der kondensierten Phosphate in verdünnter wässriger Lösung. Z Anorg Allg Chem 291:164–185

Thornton DJ, Rousseau K, McGuckin MA (2008) Structure and function of the polymeric mucins in airways mucus. Annu Rev Physiol 70:459–486

Tomas C, Brown A, Strassheim V, Elson JL, Newton J, Manning P (2017) Cellular bioenergetics is impaired in patients with chronic fatigue syndrome. PLoS One 12:e0186802

Totura AL, Whitmore A, Agnihothram S, Schäfer A, Katze MG, Heise MT, Baric RS (2015) Toll-Like receptor 3 signaling via TRIF contributes to a protective innate immune response to severe acute respiratory syndrome coronavirus infection. mBio 6:e00638-15

Tsai P (2020) Performance of masks and discussion of the inactivation of SARS-CoV-2. Eng Sci 10:1–7

Tseng YH, Yang RC, Lu TS (2020) Two hits to the renin-angiotensin system may play a key role in severe COVID-19. Kaohsiung J Med Sci 36:389–392

Tsuruta JK, Wong K, Fritz IB, Griswold MD (1990) Structural analysis of sulphated glycoprotein 2 from amino acid sequence. Relationship to clusterin and serum protein 40,40. Biochem J 268: 571–578

Ueno R, Kuno S (1987) Dextran sulphate, a potent anti-HIV agent *in vitro* having synergism with zidovudine. Lancet 1(8546):1379

Ushijima H, Rytik PG, Schäcke H, Scheffer U, Müller WEG, Schröder HC (1993) Mode of action of the anti-AIDS compound poly(I).poly(C_{12}U) (Ampligen): activator of 2′,5′-oligoadenylate synthetase and double-stranded RNA-dependent kinase. J Interferon Res 13:161–171

Vabret N, Britton GJ, Gruber C, Hegde S, Kim J, Kuksin M, Levantovsky R, Malle L, Moreira A, Park MD, Pia L, Risson E, Saffern M, Salomé B, Esai Selvan M, Spindler MP, Tan J, van der Heide V, Gregory JK, Alexandropoulos K, Bhardwaj N, Brown BD, Greenbaum B, Gümüş ZH, Homann D, Horowitz A, Kamphorst AO, Curotto de Lafaille MA, Mehandru S, Merad M, Samstein RM, Sinai Immunology Review Project (2020) Immunology of COVID-19: current state of the science. Immunity 52:910–941

Vaduganathan M, Vardeny O, Michel T, McMurray JJV, Pfeffer MA, Solomon SD (2020) Renin-angiotensin-aldosterone system inhibitors in patients with Covid-19. N Engl J Med 382:1653–1659

Van Wazer JR (ed) (1958) Phosphorus and its compounds: chemistry, vol 1. Interscience Publishers, New York

Verdecchia P, Cavallini C, Spanevello A, Angeli F (2020) The pivotal link between ACE2 deficiency and SARS-CoV-2 infection. Eur J Intern Med 76:14–20

Verhoef JJ, Barendrecht AD, Nickel KF, Dijkxhoorn K, Kenne E, Labberton L, McCarty OJ, Schiffelers R, Heijnen HF, Hendrickx AP, Schellekens H, Fens MH, de Maat S, Renné T, Maas C (2017) Polyphosphate nanoparticles on the platelet surface trigger contact system activation. Blood 129:1707–1717
Vicenzi E, Canducci F, Pinna D, Mancini N, Carletti S, Lazzarin A, Bordignon C, Poli G, Clementi M (2004) Coronaviridae and SARS-associated coronavirus strain HSR1. Emerg Infect Dis 10: 413–418
von Briesen H, Meichsner C, Andreesen R, Esser R, Schrinner E, Rübsamen-Waigmann H (1990) The polysulphated polyxylan Hoe/Bay-946 inhibits HIV replication on human monocytes/ macrophages. Res Virol 141:251–257
Wakui M, Fujimori Y, Katagiri H, Nakamura S, Kondo Y, Kuroda Y, Nakagawa T, Shimizu N, Murata M (2019) Assessment of in vitro effects of direct thrombin inhibitors and activated factor X inhibitors through clot waveform analysis. J Clin Pathol 72:244–250
Walls AC, Park YJ, Tortorici MA, Wall A, McGuire AT, Veesler D (2020) Structure, function, and antigenicity of the SARS-CoV-2 spike glycoprotein. Cell 181:281–292
Wang LF, Crameri G (2014) Emerging zoonotic viral diseases. Rev Sci Tech 33:569–581
Wang XH, Schröder HC, Müller WEG (2014) Enzymatically synthesized inorganic polymers as morphogenetically active bone scaffolds: application in regenerative medicine. Int Rev Cell Mol Biol 313:27–77
Wang XH, Schröder HC, Müller WEG (2018) Amorphous polyphosphate, a smart bioinspired nano-/bio-material for bone and cartilage regeneration: towards a new paradigm in tissue engineering. J Mat Chem B 6:2385–2412
Weiler BE, Schröder HC, Stefanovich V, Stewart D, Forrest JM, Allen LB, Bowden BJ, Kreuter MH, Voth R, Müller WEG (1990) Sulphoevernan, a polyanionic polysaccharide, and the narcissus lectin potently inhibit human immunodeficiency virus infection by binding to viral envelope protein. J Gen Virol 71:1957–1963
Westheimer FH (1987) Why nature chose phosphates. Science 235:1173–1178
Whitsett JA, Alenghat T (2015) Respiratory epithelial cells orchestrate pulmonary innate immunity. Nat Immunol 16:27–35
Wibisono Y, Fadila CR, Saiful S, Bilad MR (2020) Facile approaches of polymeric face masks reuse and reinforcements for micro-aerosol droplets and viruses filtration: a review. Polymers (Basel) 12:2516
Wickramasinghe SP, Lempart J, Merens HE, Murphy J, Huettemann P, Jakob U, Rhoades E (2019) Polyphosphate initiates tau aggregation through intra- and intermolecular scaffolding. Biophys J 117:717–728
Wong TL, Weitzer DJ (2021) Long COVID and myalgic encephalomyelitis/chronic fatigue syndrome (ME/CFS) – a systemic review and comparison of clinical presentation and symptomatology. Medicina 57:418
Wool GD, Miller JL (2021) The impact of COVID-19 disease on platelets and coagulation. Pathobiology 88:15–27
Wrapp D, Wang N, Corbett KS, Goldsmith JA, Hsieh CL, Abiona O, Graham BS, McLellan JS (2020) Cryo-EM structure of the 2019-nCoV spike in the prefusion conformation. Science 367: 1260–1263
Wright JR (2005) Immunoregulatory functions of surfactant proteins. Nat Rev Immunol 5:58–68
Xia S, Xu W, Wang Q, Wang C, Hua C, Li W, Lu L, Jiang S (2018) Peptide-based membrane fusion inhibitors targeting HCoV-229E spike protein HR1 and HR2 domains. Int J Mol Sci 19: 487
Xia S, Zhu Y, Liu M, Lan Q, Xu W, Wu Y, Ying T, Liu S, Shi Z, Jiang S, Lu L (2020) Fusion mechanism of 2019-nCoV and fusion inhibitors targeting HR1 domain in spike protein. Cell Mol Immunol 17:765–767
Xie L, Jakob U (2019) Inorganic polyphosphate, a multifunctional polyanionic protein scaffold. J Biol Chem 294:2180–2190

Xie D, Audi SH, Dash RK (2020) A size-modified Poisson-Boltzmann ion channel model in a solvent of multiple ionic species: application to voltage-dependent anion channel. J Comput Chem 41:218–230

Yan R, Zhang Y, Li Y, Xia L, Guo Y, Zhou Q (2020) Structural basis for the recognition of SARS-CoV-2 by full-length human ACE2. Science 367:1444–1448

Yang X, Wan M, Liang T, Peng M, Chen F (2017) Synthetic polyphosphate inhibits endogenous coagulation and platelet aggregation in vitro. Biomed Rep 6:57–62

Yoshida A (1955) Studies on metaphosphate. II. Heat of hydrolysis of metaphosphate extracted from yeast cells. J Biochem 42:163–168

Zeberg H, Pääbo S (2021) A genomic region associated with protection against severe COVID-19 is inherited from Neandertals. Proc Natl Acad Sci USA 118:e2026309118

Zhang R, Jha BK, Ogden KM, Dong B, Zhao L, Elliott R, Patton JT, Silverman RH, Weiss SR (2013) Homologous 2′,5′-phosphodiesterases from disparate RNA viruses antagonize antiviral innate immunity. Proc Natl Acad Sci USA 110:13114–13119

Zhang YH, Zhang YH, Dong XF, Hao QQ, Zhou XM, Yu QT, Li SY, Chen X, Tengbeh AF, Dong B, Zhang Y (2015) ACE2 and Ang-(1-7) protect endothelial cell function and prevent early atherosclerosis by inhibiting inflammatory response. Inflamm Res 64:253–260

Zhang S, Liu Y, Wang X, Yang L, Li H, Wang Y, Liu M, Zhao X, Xie Y, Yang Y, Zhang S, Fan Z, Dong J, Yuan Z, Ding Z, Zhang Y, Hu L (2020) SARS-CoV-2 binds platelet ACE2 to enhance thrombosis in COVID-19. J Hematol Oncol 13:120

Zheng S, Fan J, Yu F, Feng B, Lou B, Zou Q, Xie G, Lin S, Wang R, Yang X, Chen W, Wang Q, Zhang D, Liu Y, Gong R, Ma Z, Lu S, Xiao Y, Gu Y, Zhang J, Yao H, Xu K, Lu X, Wei G, Zhou J, Fang Q, Cai H, Qiu Y, Sheng J, Chen Y, Liang T (2020) Viral load dynamics and disease severity in patients infected with SARS-CoV-2 in Zhejiang province, China, January-March 2020: retrospective cohort study. BMJ 369:m1443

Zhou S, Butler-Laporte G, Nakanishi T, Morrison DR, Afilalo J, Afilalo M, Laurent L, Pietzner M, Kerrison N, Zhao K, Brunet-Ratnasingham E, Henry D, Kimchi N, Afrasiabi Z, Rezk N, Bouab M, Petitjean L, Guzman C, Xue X, Tselios C, Vulesevic B, Adeleye O, Abdullah T, Almamlouk N, Chen Y, Chassé M, Durand M, Paterson C, Normark J, Frithiof R, Lipcsey M, Hultström M, Greenwood CMT, Zeberg H, Langenberg C, Thysell E, Pollak M, Mooser V, Forgetta V, Kaufmann DE, Richards JB (2021) A Neanderthal OAS1 isoform protects individuals of European ancestry against COVID-19 susceptibility and severity. Nat Med 27:659–667

MIX
Papier aus verantwortungsvollen Quellen
Paper from responsible sources
FSC® C105338

If you have any concerns about our products,
you can contact us on
ProductSafety@springernature.com

In case Publisher is established outside the EU,
the EU authorized representative is:
Springer Nature Customer Service Center GmbH
Europaplatz 3, 69115 Heidelberg, Germany

Printed by Libri Plureos GmbH
in Hamburg, Germany